Backscattered Scanning Electron Microscopy and Image Analysis of Sediments and Sedimentary Rocks

Backscattered scanning electron microscopy (BSE) reveals the minerals, textures, and fabrics of sediments and rocks in much greater detail than is possible with conventional optical microscopy. *Backscattered Scanning Electron Microscopy* provides a concise summary of the BSE technique. This comprehensive guide uses abundant images to illustrate the type of information BSE yields and the application of the technique to the study of sediments and sedimentary rocks.

The authors review the use of this petrographic technique on all the major sedimentary rock types, including sediment grains, sandstones, shales, carbonate rocks, rock varnish, and glauconite. They also describe image analysis techniques that allow quantification of backscattered scanning electron microscope images and illustrate the potential applications of these methods.

Heavily illustrated and lucidly written, this book provides researchers and graduate students with the most current research on this important geological tool.

Backscattered Scanning Electron Microscopy and Image Analysis of Sediments and Sedimentary Rocks

DAVID H. KRINSLEY
KENNETH PYE
SAM BOGGS, JR.
N. KEITH TOVEY

CAMBRIDGE
UNIVERSITY PRESS

CAMBRIDGE UNIVERSITY PRESS
Cambridge, New York, Melbourne, Madrid, Cape Town, Singapore, São Paulo

Cambridge University Press
The Edinburgh Building, Cambridge CB2 2RU, UK

Published in the United States of America by Cambridge University Press, New York

www.cambridge.org
Information on this title: www.cambridge.org/9780521453462

First published 1998
This digitally printed first paperback version 2005

A catalogue record for this publication is available from the British Library

Library of Congress Cataloguing in Publication data
Backscattered scanning electron microscopy and image analysis of
 sediments and sedimentary rocks / David H. Krinsley . . . [et al.].
    p. cm.
  Includes bibliographical references (p. –) and index.
  ISBN 0-521-45346-1
  1. Rocks, Sedimentary – Analysis. 2. Sediments (Geology) –
Analysis. 3. Scanning electron microscopy. I. Krinsley, David
H. (David Henry), 1927– .
  QE471.B26 1998   97-18011
  552´.5´028 – dc21   CIP

ISBN-13 978-0-521-45346-2 hardback
ISBN-10 0-521-45346-1 hardback

ISBN-13 978-0-521-01974-3 paperback
ISBN-10 0-521-01974-5 paperback

Contents

Preface   vii

Acknowledgments   ix

1   Introduction   1

2   The nature of backscattered
    scanning electron images   4

3   Sediment grains, weathering, and early
    diagenetic phenomena   25

4   Sandstones   52

5   Shales   73

6   Carbonates   98

7   Desert (rock) varnish   119

8   Glauconite   131

9   Image analysis   145

    Bibliography   173

    Index   191

Preface

This book focuses on the study of sedimentary rock thin sections using images obtained from the scanning electron microscope (SEM) in the backscattered electron mode (BSE). About half the text consists of description and analysis, and the rest SEM micrographs. Until now, there has not been a book available that provides BSE information on the various sediments and sedimentary rock types. We hope the book will provide an analytical tool for sedimentologists and others who may profit from a reference volume for use in their studies.

The earliest known work concerning the construction of an SEM was published in 1938. Much of the subsequent development was done in the engineering department of the University of Cambridge between 1948 and 1965. The first BSE pictures with good resolution were produced in the 1970s. Today, BSE photographs of sediments and sedimentary rocks appear regularly in the major sedimentological journals.

The basic components of an SEM are the lens system, electron gun, electron collector, visual and recording cathode ray tubes (CRTs), and the electronics associated with them. BSE is simply one of the operational modes. Electrons from the primary beam of an SEM strike the atoms of a sample and are reflected or scattered (backscattered) out of the material at high angles. These electrons are collected and used to produce a gray-level photograph that contains both compositional (atomic number) and topographic information.

Because an image in the BSE mode has a much greater resolution than can be produced with the ordinary light microscope (in addition to atomic number information), BSE has been used in sedimentology and associated fields to provide textural data that are somewhat similar to those obtained with the light petrographic microscope. For instance, when a thin section is viewed using this technique, size, shape, distribution, and composition of sedimentary grains can be observed at magnifications of from 25 to 20,000 times, with resolutions at least 10 times those of the light microscope. The mineralogy and porosity of very fine particles can be obtained; additionally, information about diagenetic alteration of very fine particles, including cementation and replacement, are routinely acquired. Thus the fine-grained fraction of sediments and sedimentary rocks can be examined in great detail, and new insights made available to the microscopist.

The book is divided into four parts. Chapters 1, 2, and 3 are introductory; they discuss the nature of BSE images, and BSE's general application to weathering and early diagenesis of sedimentary grains. Chapters 4, 5, and 6 are concerned with sandstones, shales, and carbonates. Chapters 7 and 8 involve specific examples taken from our work on desert varnish and glauconite. Chapter 9 discusses image analysis of BSE photographs.

SEM/BSE instruments are commonly available to researchers in academia and industry. As these microscopes become smaller and less expensive, the SEM in its various modes will become widely used by undergraduates, graduates, and researchers wishing to obtain submicroscopic information on rocks and minerals. The value of BSE is limited only by the imagination of the researcher.

Acknowledgments

Numerous individuals and organizations have assisted the research from which the examples used in this book have been drawn. Financial support was provided by NATO, the Royal Society, the U.K. Natural Environment Research Council, the Petroleum Research Fund of the American Chemical Society, the U.S. Department of Energy, ARCO, British Petroleum, Schlumberger Cambridge Research, and the Ocean Drilling Program, Texas A&M University. We particularly wish to thank J. Burton, A. Cross, R. Dorn, H. Dypvik, P. Isles, J. Jack, A. Kearsley, D. Newling, W. Porter, I. Pryde, A. Seyedolali, M. B. Shaffer, K. Smith, P. Trusty, and J. Watkins for assistance with specimen preparation, maintenance of scanning electron microscope equipment, and production of illustrations.

1

Introduction

During the past two to three decades, the scanning electron microscope (SEM) has become established as an essential tool in the study of sedimentary rocks, sediments, and soils. It provides a useful complement to the traditional role of the petrographic microscope, which became popular after the pioneering work of Henry Clifton Sorby in the second half of the nineteenth century (Sorby, 1877a,b, 1878). The modern SEM is a versatile analytical work station, capable of providing several different types of images, quantitative data relating to porosity and rock composition, and information about the crystallographic structure of individual minerals. The various images and chemical data can be processed by on-line computers or stored on disk for later analysis. Computer control can also be used to drive the electron beam, facilitating routine, automated analysis of samples (e.g., Minnis, 1984; Cook and Parker, 1988; Habesch, 1990; Ehrlich et al., 1991; Dijkshoorn and Fens, 1992).

Prior to the early 1980s, most SEM work in geology utilized the secondary electron (SE) mode to examine fine surface textural detail on sediment grains, fossils, and fracture surfaces of rocks (e.g., Krinsley and Doornkamp, 1973; Smart and Tovey, 1981; Welton, 1984; Trewin, 1988). This technique provides a great deal of useful textural information relating to sediment provenance, diagenesis, weathering history, and geotechnical properties. Its value is limited to some extent; however, because often it is not possible to distinguish among minerals that have similar morphologies, the textural relationships between different mineral phases are not always clear, and quantification of the sediment fabric is typically difficult.

The capability of using backscattered electrons to produce images on the basis of compositional contrast has existed since the early 1960s, but Kiss and Briskies (1976) and Robinson and Nickel (1979) were among the first workers to draw attention to the potential value of this technique in the context of mineralogical and wider geological applications. The 1970s saw rapid improvements in the design of BSE detectors and consequently in image quality. During the early 1980s, the value of the BSE imaging techniques in the study of sediments, soils, and rocks was demonstrated by several authors (e.g., Hall and Lloyd, 1981; Bisdom and Thiel, 1981; Jongerius and Bisdom, 1981; Krinsley et al., 1983; Pye and Krinsley, 1983; 1984; White et al., 1984). The technique proved especially useful for study of fine-grained materials, such as clay or shale, about which little could be said using conventional optical microscopy. BSE microscopy has now become a standard petrographic technique that is used to provide information about the nature of intergranular cements, diagenetic history, fossil preservation, metamorphism, and rock weathering (Krinsley and Manley, 1989). There is, however, no generally available summary of the technique and its applications in sedimentary geology. The principal purpose of this book is to provide such an illustrative summary, which will be a useful introduction for advanced undergraduates, research students and more established research workers who wish to broaden their expertise in sedimentary petrology. We have not attempted to produce a comprehensive atlas of sediments and sedimentary rocks because the range of textures, minerals, and fabrics that could be illustrated is almost infinite; rather, the micrographs and associated text are intended to illustrate the *types* of information that can be obtained by using BSE microscopy. Increasingly, BSE images are being used as the basis for the quantitative analysis of the mineral composition and fabric of sediments, soils, and sedimentary rocks (e.g., Dilks and Graham, 1984, 1985; Ehrlich et al., 1984; Evans et al., 1994; Tovey and Krinsley, 1991, 1992; Tovey and Hounslow, 1995). Consequently, the final chapter of this book considers the techniques available to quantify BSE images and illustrates some of the potential applications. Extensive referencing has been employed throughout the book so that the interested reader can easily find further information relating to points of particular concern.

2

The Nature of Backscattered Scanning Electron Images

SPECIMEN–ELECTRON BEAM INTERACTIONS

When a beam of primary electrons emitted by a filament in an electron gun hits the surface of a specimen, excitation by the beam causes the emission of secondary electrons, backscattered (reflected primary) electrons, Auger electrons, photons, characteristic X-rays, continuum X-rays, and heat (Fig. 2.1). These signals are produced from a specific emission volume within the specimen, the size of which depends upon the primary electron beam energy (E_o) and the average atomic number (Z) of the specimen. The incident primary electrons undergo two types of scattering upon entering the specimen: **elastic scattering** and **inelastic scattering**. During elastic scattering, the incident electrons interact with the nuclei of atoms in the specimen, causing a significant deflation of the trajectories, but they lose little energy. Two types of elastic scattering can occur. Rutherford scattering, which occurs in the Coulomb field of the nucleus, is a single scattering event that results in a large change in trajectory direction, in some cases exceeding 90°. Multiple scattering comprises several small scattering events. A proportion of the elastically scattered electrons is eventually directed back out of the specimen as **backscattered electrons**. The remainder diffuse through the sample in a random manner and are eventually absorbed. During inelastic scattering, incident electrons may either (1) interact with the nuclei of atoms in the specimen, losing energy in the Coulomb field and emitting white or continuum X-ray radiation or (2) collide with loosely bound electrons in the specimen, causing the latter to be ejected as **secondary electrons.** Because secondary electrons have low energies (typically < 50 eV), they are unlikely to escape from the specimen if produced at a depth greater than approximately 50 Å below the surface. These secondary electrons are used for conventional imaging in the scanning electron microscope (SEM) because they provide information about fine surface detail. If the secondary electrons recombine with holes, or vacancies, formed during the scattering process, photons are produced that have wavelengths in the visible or near-infra-red ranges. These photons can be collected and used for imaging in the cathodoluminescence mode (e.g., Autrata et al., 1992; Saparin and Obyden, 1993).

The nature of the scattering events induced by the incident electron beam is strongly controlled by the average atomic number of the specimen. In the case of low atomic number materials, little scattering takes place near the surface; consequently, only a few electrons are scattered through

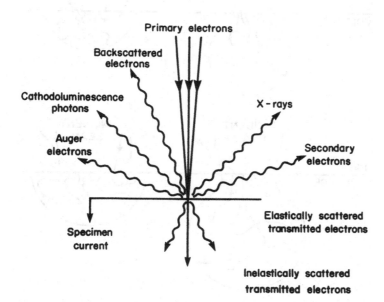

Figure 2.1. Signals resulting from interaction between a primary electron beam and specimen.

large angles and emerge as backscattered electrons. Most of the incident electrons penetrate deeply into the specimen and are eventually absorbed. With high atomic number materials there is much greater scattering close to the specimen surface; consequently, a larger number of incident electrons are reflected by more than 90° and emerge as backscattered electrons. Figure 2.2 shows that, for the same energy, incident electrons penetrate more deeply into a low atomic number material, and the electron diffusion envelope is more pear-shaped than is the case for high atomic number material. Penetration is less in high atomic number material because a large number of scattering events causes random diffusion of electrons close to the specimen surface.

The amount of electron backscattering is indicated by the backscattering coefficient (η), which is defined as the fraction of incident electrons that do not remain in the specimen:

Figure 2.2. Section through a specimen illustrating the variation of electron scattering with incident beam voltage and atomic number (after Duncumb and Shields, 1963).

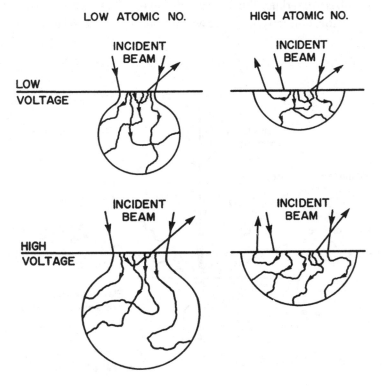

$$\eta = \frac{\text{number of backscattered electrons}}{\text{number of incident electrons}}$$

It has proved difficult to predict η accurately owing to the complexity of the scattering processes involved, so heavy reliance has been placed on experimental data. Bishop (1966) found experimentally that, for atomic numbers less than 47, η increases slightly with incident beam energy (E_o) over the energy range 10–30 keV, but for higher atomic numbers η decreases slightly with E_o (Fig. 2.3). For electron beam energies of less than 10 keV, η varies markedly with E_o and is not a simple function of atomic number (Z) (Darlington and Cosslett, 1972). A strong dependence of η on atomic number has been demonstrated by several authors (e.g., Bishop, 1966; Heinrich, 1966; Wittry, 1966). For atomic numbers up to 30, η increases approximately linearly with Z, but for Z greater than 30 the curve flattens out, reaching a value of about $\eta = 0.5$ for $Z = 90$ (Fig. 2.4). In the case of pure ele-

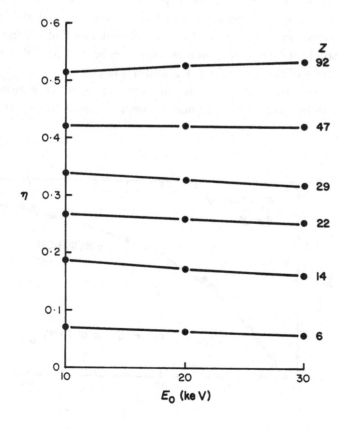

Figure 2.3. Dependence of backscattering electron coefficient (η) on incident electron energy (E_o [keV]) (after Bishop, 1966).

ments ($Z > 10$) the relationship between η and Z is given by the following equation (Heinrich, 1966):

$$\eta = (\ln Z/6) - 0.25$$

Reed (1975) calculated electron backscattering coefficients for all the elements up to $Z = 92$ on the basis of Bishop's (1966) experimental data for 16 elements (Table 2.1 and Fig. 2.4). Each value is based on an average of Bishop's data for 10 keV and 30 keV.

When the specimen is composed of a homogeneous mixture of several elements, a simple rule of mixtures based on the weight fractions of each component has been found to apply (Heinrich, 1964):

$$\eta_{compound} = \sum_{i=1}^{n} C_i \eta_i$$

where C_i is the weight fraction of each element, η_i is the elemental backscattering coefficient, and n is the number of elements. Values of Z and η for a number of common minerals of ideal formula present in rocks and related materials are shown in Table 2.2. In the case of minerals that belong to solid solution series, such as pyroxenes and amphiboles, the range of possible values of Z and η is large.

Although some manufacturers of detectors claim atomic number contrast resolution of better than 0.003 Z, most

Figure 2.4. Comparison of backscattered electron coefficient (η) and secondary electron coefficient (δ) as a function of atomic number (Z); data of Heinrich (1966) and Wittry (1966) (after Goldstein et al., 1981).

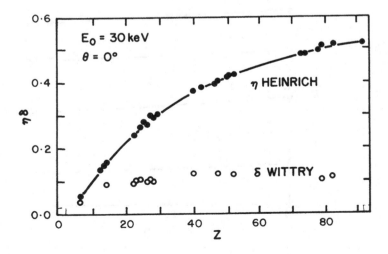

available detectors are routinely able to resolve differences in Z of about 0.1. Consequently, it is not always possible to differentiate between some minerals – for example, quartz and dolomite or albite – on the basis of gray level. On the other hand, many common minerals can be readily distin-

TABLE 2.1

Electron backscattering coefficient for pure elements

Z	η	Z	η	Z	η	Z	η
1	0.010	24	0.282	47	0.412	70	0.482
2	0.021	25	0.291	48	0.415	71	0.485
3	0.031	26	0.300	49	0.419	72	0.487
4	0.043	27	0.309	50	0.423	73	0.490
5	0.054	28	0.317	51	0.426	74	0.492
6	0.066	29	0.325	52	0.430	75	0.495
7	0.079	30	0.332	53	0.433	76	0.498
8	0.093	31	0.339	54	0.436	77	0.501
9	0.107	32	0.345	55	0.439	78	0.504
10	0.121	33	0.350	56	0.443	79	0.507
11	0.136	34	0.355	57	0.446	80	0.509
12	0.150	35	0.360	58	0.449	81	0.510
13	0.164	36	0.365	59	0.452	82	0.512
14	0.176	37	0.370	60	0.455	83	0.513
15	0.189	38	0.375	61	0.458	84	0.515
16	0.201	39	0.379	62	0.461	85	0.516
17	0.212	40	0.384	63	0.463	86	0.517
18	0.223	41	0.388	64	0.466	87	0.518
19	0.233	42	0.392	65	0.469	88	0.519
20	0.243	43	0.396	66	0.472	89	0.520
21	0.253	44	0.400	67	0.474	90	0.521
22	0.263	45	0.404	68	0.477	91	0.522
23	0.273	46	0.408	69	0.479	92	0.523

Source: Reed, 1975, p. 223.

guished. In the case of phyllosilicates, relatively slight variations in composition, including the structural water content, can be detected within individual particles and aggregates (Pye and Krinsley, 1983; Huggett, 1984, 1986; White et al., 1984, 1985). For example, a 0.5% weight change in the FeO content or a 1% change in MgO content of a chlorite will alter Z by approximately 0.1; a 1% change in the K_2O content will achieve the same result in the case of illite (White et al., 1984).

The value of η also varies with tilt angle (θ) of the specimen relative to the incident beam. If θ is increased, the size

TABLE 2.2

Mean atomic number ($\bar{Z}$) and calculated backscattering coefficients (η) for some minerals of ideal formula commonly found in sedimentary rocks

Mineral	Formula	$\bar{Z}$	η
Mg chlorite	$Mg_5Al_2Si_3O_{10}(OH)_8$	10.17	0.124
Kaolinite	$Al_4Si_4O_{10}(OH)_8$	10.24	0.125
NaMontmorillonite	$Na_{0.33}(Al_{1.67}Mg_{0.33})Si_4O_{10}(OH)_2\cdot H_2O$	10.40	0.127
CaMontmorillonite	$Ca_{0.17}(Al_{1.67}Mg_{0.33})Si_4O_{10}(OH)_2\cdot 2H_2O$	10.41	0.127
Albite	$NaAlSi_3O_8$	10.71	0.132
Quartz	SiO_2	10.80	0.132
Dolomite	$CaMg(CO_3)_2$	10.87	0.133
Illite	$K_{1.5}Al_{5.5}Si_{6.5}O_{20}(OH)_4$	11.16	0.136
Muscovite	$K_2Al_6Si_6O_{20}(OH)_4$	11.33	0.138
Orthoclase	$KAlSi_3O_8$	11.85	0.145
Calcite	$CaCO_3$	12.57	0.150
Biotite	$K_2(Mg_2Fe_4)Al_2Si_6O_{20}(OH)_4$	14.59	0.174
Fe chlorite	$Fe_5Al_2Si_3O_{10}(OH)_8$	16.05	0.188
Siderite	$FeCO_3$	16.47	0.190
Rutile	TiO_2	16.93	0.195
Goethite	$FeO\cdot OH$	19.23	0.222
Hematite	Fe_2O_3	20.59	0.238
Pyrite	FeS_2	20.65	0.247

of the electron interaction volume decreases and the opportunity for backscattering events is increased (Goldstein et al., 1981). Newbury et al. (1973) showed that η increases slowly with an increase in θ up to 20° and rapidly at higher tilt angles (Fig. 2.5). Because η for all elements tends toward the same value at high tilt angles, atomic number contrast is reduced.

The angular distribution of backscattered electrons is also dependent upon the specimen tilt angle. At normal incidence ($\theta = 0$), the distribution follows a cosine law:

$$\eta(\Phi) = \eta'\cos\Phi$$

where Φ is the angle between the surface normal and the direction of measurement and η' is the value of η along the surface normal.

With increasing tilt, the distribution becomes progressively more peaked in the forward scattering direction. This directionality of backscattering from tilted surfaces creates a trajectory component that provides the basis for topographic contrast when rough surfaces are imaged in the backscattered electron (BSE) mode.

The emission of secondary electrons from a specimen is much less dependent on atomic number than is the emission of backscattered electrons (Wittry, 1966; Fig. 2.4); however, some atomic number contrast is observed when flat, polished surfaces are examined in the SEM using the secondary electron (SE) mode (e.g., Krinsley et al., 1983). This contrast

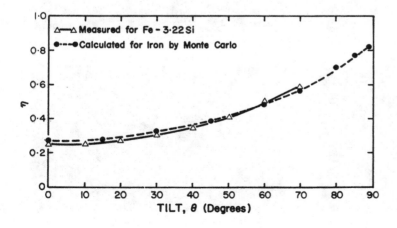

Figure 2.5. Backscatter coefficient (η) as a function of beam tilt, based on Monte Carlo calculations and empirical measurements (after Newbury et al., 1973).

occurs because a significant proportion of secondary electrons is generated by backscattered electrons, the exact proportion varying with Z from 0.18 in the case of carbon to 1.5 in the case of gold (Goldstein et al., 1981, p. 91).

The SE and BSE images obtained from a polished thin section of sandstone are compared in Figure 2.6. The much greater atomic number contrast observed in the BSE image is clearly evident. A few instances have been reported in which greater compositional contrast has been observed in the SE image than in the BSE image (e.g., Sawyer and Page, 1978). This phenomenon appears to occur primarily in semiconductors owing to the presence of trace impurities, which can modify the electron acceptor levels in the electronic band structure without affecting Z and η.

A further difference between the SE and BSE images in Figure 2.6 is that excessive edge brightness is apparent in the SE image but is absent in the BSE image. Excessive edge brightness arises in the SE image because polishing of the specimen has not produced a perfectly flat surface owing to hardness differences; consequently, there is a contribution

Figure 2.6. Four different images of the same area of a quartz sandstone from the southern North Sea: (a) 100% secondary electron image, showing good resolution morphological information but low atomic number contrast; (b) 50% secondary electron and 50% backscattered electron image; (c) 75% backscattered and 25% secondary electron image; (d) 100% backscattered electron image, showing high atomic number contrast but slight loss of edge definition. Carbon-coated polished sections, signal mixing at 30 keV using a Phillips multifunction detector. Scale bars = 30 μm.

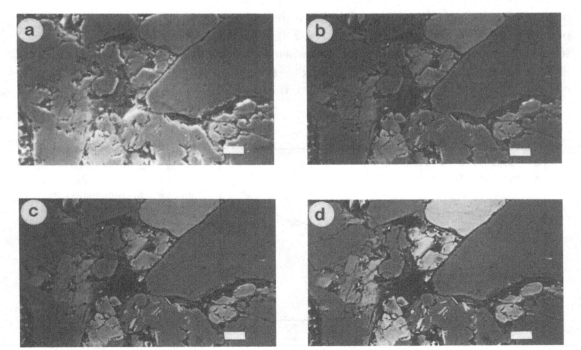

to the SE signal from forward-scattered primary electrons that penetrate the grain edges (Robinson and George, 1978; Wells, 1978, 1986; Robinson, 1988). In addition, specimen charging is often much more of a problem in SE images than in BSE images. The combination of edge effects and charging can cause serious loss of detail in fine-grained specimens (Krinsley et al., 1983). Secondary electrons have an average energy of only 1–2 eV; if a specimen is insufficiently conducting, the buildup of charges on the order of 2 volts is sufficient to cause charging artifacts. Backscattered electrons have an average energy of 50–90% of the primary electrons; thus, at high accelerating voltages (>10 kV), a charge buildup of several thousand volts is required before charging effects appear (Robinson, 1988). Consequently, it is often possible to generate images in the BSE mode from uncoated specimens, whereas it is almost always necessary to coat geological specimens with gold or carbon before images can be generated in the SE mode. In some applications, such as the investigation of fossil preservation as carbon films in shales, the necessity to coat the specimens imposes a serious constraint because the conductive surface coating may totally obscure the feature of interest.

The spatial resolution attainable in BSE images is usually inferior to that obtainable in SE images, for two main reasons. First, backscattered electrons emerge from a greater depth range in the specimen than do secondary electrons. The maximum emergence depth for backscattered electrons shows an inverse relationship with Z and a positive relationship with E_o (Murata et al., 1968; Fig. 2.7). In the case of typical silicate minerals examined at operating voltages of 15–20 keV, backscattered electrons may emerge from depths as great as 1 μm below the specimen surface, although the majority emerge from shallower depths. Second, some backscattered electrons can emerge from the specimen surface at a significant distance from the incident beam impact area. For a beam normally incident on a flat surface, the distribution of backscattered electrons is symmetrical and sharply

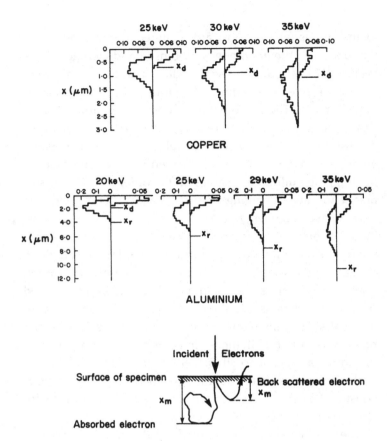

Figure 2.7. Distribution of maximum penetration depth for backscattered electrons (right-hand side) and absorbed electrons (left-hand side) for copper and aluminium targets (data of Murata et al., 1968). Both x and y axes are in microns. x_r is the maximum range of electrons, x_d is the calculated depth of complete diffusion of electrons, and x_m is maximum penetration depth of electrons.

Figure 2.8. Calculated surface distribution of backscattered electrons for copper at 20 keV (after Murata, 1974).

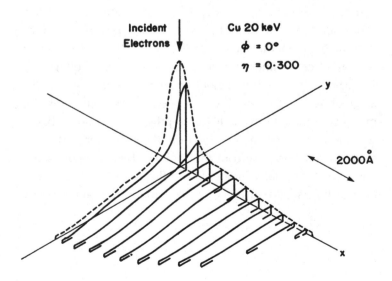

peaked at the beam impact point (Fig. 2.8). Monte Carlo simulation by Murata (1973, 1974) showed that, in the case of a copper target and operating voltage of 20 keV, backscattered electrons escape up to 9 μm from the center of the beam impact area. For such high atomic number targets, the diameter of the BSE spatial distribution is smaller, and the central peak higher, than is the case for low atomic number targets such as aluminum. By comparison, the distribution of secondary electrons shows much greater spatial concentration because the majority of such electrons are generated by incident electrons and because low-energy secondary electrons can escape from the surface only if they have a short travel distance through the specimen (Fig. 2.9). Although the boundaries between mineral grains in a geological specimen are normally sharp at the atomic level, in higher-magnification BSE images they appear diffuse owing to the fact that backscattered electrons may cross the boundary from one grain to another before emerging at the surface. Under ideal condi-

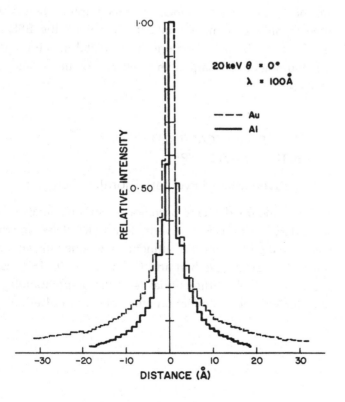

Figure 2.9. Lateral distribution of secondary electrons from gold and aluminium at 20 keV (after Murata, 1973).

tions of high atomic number contrast (e.g., gold particles on a carbon substrate), however, a spatial resolution of better than 0.01 mm is obtainable with some high signal to noise ratio detectors. In the case of typical sedimentary rocks, useful images at magnifications up to 10,000 × can normally be obtained by using operating voltages between 10 and 30 keV.

For routine investigation of geological samples, in which both images of the sample and qualitative or quantitative analysis of individual mineral components using energy-dispersive or wavelength-dispersive spectrometry are required, it is customary to use an operating voltage in the range of 15–30 keV (most commonly 20 keV). Using a high beam voltage and large spot size (high beam current), good atomic number contrast can be obtained; however, problems with spatial resolution may arise owing to excessive beam penetration and electron scattering. Some of these problems may be avoided by using a smaller spot size or, with some instruments, a low operating voltage (< 5 keV) (Lanteri et al., 1988). Better image definition may also be obtained by tilting the specimen to enhance the topographic contrast relative to atomic number contrast, by mixing the BSE and SE signals, or by using some form of digital image processing equipment to sharpen the "edges" of the image (see Chapter 9).

TYPES OF BACKSCATTERED ELECTRON DETECTORS

Negatively Biased Everhart-Thornley Detector

The standard detector used for generating images with secondary electrons is the Everhart-Thornley detector, which was developed in the department of engineering at Cambridge University (Everhart and Thornley, 1960). This instrument is a highly directional scintillator–photomultiplier detector with good collection efficiency and noise-free amplification (Fig. 2.10a). Because secondary electrons have low energy, the detector can be placed well away from the specimen and surrounded by a metal cage to which a biasing voltage is applied to attract the electrons toward the detector. With this detector, the image is actually formed by a combination of secondary and backscattered electrons, and it is possible to vary the relative contribution of each to the final image by altering the bias on the copper grid around the detector. If the bias is reversed, creating a negative charge, secondary electrons can effectively be excluded from the detector. Because the Everhart-Thornley detector has a low solid angle with respect to the specimen stage, however, it has a low collection efficiency for backscattered electrons, which have higher energy (typically up to 10 keV) and are unaffected by the biasing voltage. Therefore, the BSE image obtained by using the Everhart-Thornley detector has a poor signal to noise ratio.

Wide-Angle Scintillator–Photomultiplier Detectors

To improve the solid angle for more efficient BSE collection, a large scintillator must be placed close to the specimen (Wells, 1970, 1977; Schur et al., 1974; Robinson, 1973, 1975). Schur et al. (1974) described a large area scintillator detector connected to a light pipe (Fig. 2.10b). The specimen is tilted toward the detector; this directionality of the detector tends to enhance topographic contrast effects. Robinson (1973) developed a highly efficient nondirectional hemispherical scintillator detector, which virtually surrounded the specimen. The prototype was not suitable for examination of large specimens or simultaneous X-ray microanalysis, but subsequent commercial modifications of the design are more compact and less intrusive (Fig. 2.10c). A flat detector head is located directly above the sample and has a central hole through which the incident beam passes. Robinson-type detectors provide excellent imaging at television rates,

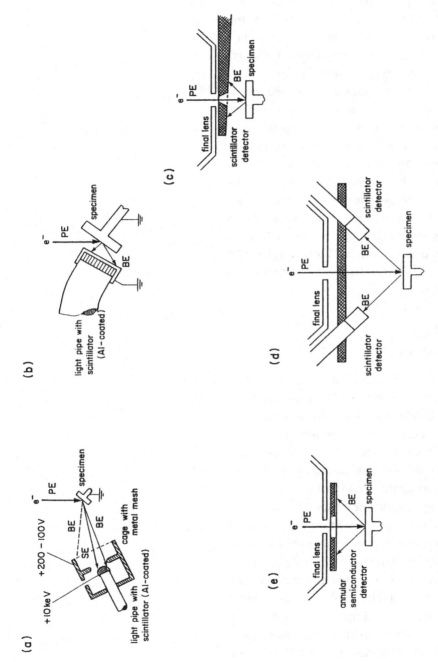

Figure 2.10. Schematic representation of five detector types used for backscattered electron imaging: (a) Everhart-Thornley detector (Everhart and Thornley, 1960); (b) directional scintillator-photomultiplier detector (Schur et al., 1974); (c) Robinson scintillator detector (Robinson, 1975); (d) multifunction scintillator-photomultiplier detector (Jackman, 1980); (e) annular solid-state detector (Wolf and Everhart, 1969). e⁻ = electrons; BE = backscattered electrons; SE = secondary electrons; PE = primary electrons.

have a high signal to noise ratio, and give good atomic number contrast (Robinson, 1980, 1988).

Multiple Scintillator Arrays

In this type of system, exemplified by the Phillips multifunction detector system, one or two pairs of phosphor-coated scintillator detectors are mounted below the final lens and are linked by fiber-optic cables to photomultiplier–preamplifier combinations (Jackman, 1980; Reimer and Volbert, 1980; Pye and Windsor-Martin, 1983; Fig. 2.10d, e). The two signals reaching each photomultiplier can be separated and used singly or in sum. This type of system gives a good solid angle for efficient backscattered electron collection, high signal to noise ratio, and the versatility provided by signal mixing capability (Volbert, 1982).

Reimer and Riepenhausen (1985) and Hejna (1987) described variants of a ring scintillator BSE detector that surrounds the specimen and accepts electrons emitted in a specific range of zenith angles. In the case of untilted samples, this type of detector collects electrons that emerge from the surface at low angles. The images obtained show suppressed atomic number contrast, improved spatial resolution, and reduced shadowing effect compared with images obtained when a solid-state detector is placed to one side of the specimen.

Solid-State Detectors

Several variants of this type of detector have been developed for different purposes (e.g., Kimoto et al., 1966; Stephen et al., 1975; Lin and Becker, 1975; Gedke et al., 1978; Hejna et al., 1985; Radzimski, 1987). Those most widely present on commercial SEMs consist of either a pair of (commonly rectangular) diodes that are placed above the specimen stage on either side of the polepiece (Gedke et al., 1978) or an annular silicon diode, which is mounted immediately below the final lens concentric with the electron beam (Wolf and Everhart, 1969; Fig. 2.10e). The annular diode commonly has four separate elements, each connected to an amplifier, which can be operated independently or in combination. The BSE signal received by each quadrant is affected slightly differently by the surface topography of the specimen. Adding the signals to opposite sectors gives an A + B image in which the contrast is primarily due to atomic number differences, whereas subtracting the two signals gives an A − B image that is primarily topographical (Kimoto et al., 1966). This facility can be useful if it is necessary to separate true topography from pseudotopographic contrast generated by differential scattering effects at crystal boundaries. Early solid-state detectors did not produce good TV-rate images (Robinson and Nickel, 1983), but this problem has been largely overcome with later versions (Hall and Lloyd, 1983). Modern solid-state detectors have signal to noise characteristics comparable with scintillator–photomultiplier detectors and have the advantage of being compact and relatively inexpensive. However, they are somewhat fragile and subject to breakage unless care is taken to ensure adequate clearance between a sample and the detector.

Low-Voltage Backscattered Electron Detectors

A number of detectors have been developed specifically to assist image acquisition from sensitive and uncoated samples (mainly biological specimens and nonconducting materials) at low-kV and small probe currents. An example is the microchannel plate detector, which gives reasonable-resolution BSE and SE images in the 1–5 keV range (Joy and Joy, 1993). The microchannel plate detector consists of an array of electron multiplier tubes, each a few micrometers in diameter, assembled into a closely packed array to form a disc about 2 cm in diameter. Backscattered electrons that strike the front face of the detector produce low-energy electrons that are accelerated down the tube and collected by a biased anode connected to an external amplifier. These devices function efficiently down to beam energies below 1 keV and remain usefully sensitive for energies above 30 keV. By applying a positive bias on the screen on the entrance side of the detector, incoming secondary electrons can be accelerated sufficiently to obtain a topographic image.

SPECIMEN PREPARATION

Water Removal, Embedding, and Impregnation

The quality of the SEM images ultimately obtained is heavily dependent upon sample preparation. A flat, well-polished surface is the single most important requirement to obtain high-quality atomic number contrast images. Some hard-rock samples can be cut and polished in a relatively straightforward manner; more sensitive or fragile samples require careful pretreatment and handling to ensure that plucking, smearing, selective dissolution, or excessive shrinkage does not occur. In the case of hard, well-cemented rocks with low porosity, polished thin sections that are prepared for reflected light microscopy or electron microprobe analysis can be used quite satisfactorily for BSE examination.

Friable, soft rocks and loose sediments require pretreatment to remove water or other impurities and to give the material strength before cutting and polishing. If optical microscope examination is not required, it is frequently easier to prepare a polished rock chip or thick section than a conventional thin section of 30 μm thickness. A variety of procedures for resin impregnation have been described in the literature. If a sample contains significant amounts of moisture and it is necessary to avoid changes in texture or mineral composition owing to shrinkage and dehydration, the water can be replaced gradually using mixtures of acetone and water (e.g., Bailey and Blackson, 1984). After the sample has been immersed in 100% acetone for at least 12 hours, the acetone is allowed to evaporate; the sample can then be infiltrated at ambient pressure with a low-viscosity epoxy resin. A very wide range of embedding resins is available commercially. Many are hazardous and require great care during handling (see Causton, 1988, for a brief review).

Murphy (1982) compared the effects of oven-drying, freeze-drying, and acetone replacement on the results of porosity determinations using a Quantimet 720 image analyzer. He found in the case of both oven-drying and freeze-drying that major macroshrinkage caused a slight decrease in the number and lengths of most planar pores and a decrease in the size and number of intraaggregate pores. The loss of pore spaces subsequently made it more difficult to achieve satisfactory resin impregnation. Clay-rich samples in which the water was replaced by acetone showed no measurable macroshrinkage and impregnated well with resin.

To fit into most automatic grinding machines, impregnated samples must be less than 25 mm long and 5 mm thick. It may, therefore, be necessary to trim an initially larger block of impregnated sediment or rock and to reset it in a suitable-sized mould to which further epoxy is added.

To achieve maximum penetration of epoxy into the pore spaces, it is advisable to carry out the impregnation operation under vacuum. A low-viscosity, low-temperature curing epoxy, or one that can be cured using ultraviolet light, should be used for heat-sensitive specimens. For satisfactory results, the epoxy should display good adherence to the rock matrix, have a very low vapor pressure so that the likelihood of boiling under vacuum is reduced, and show minimum shrinkage during curing. Several commercially available alternatives have been successfully used, including Spurr epoxy (Spurr, 1969; Jim, 1986) and Epofix (Dijkshoorn and Fens, 1992).

If a sample consists of loose grains or small chips (e.g., well cuttings), these can be set in a resin block before cutting and polishing. Alternatively, a number of chips can be individually oriented and glued onto a glass disc or slide suitable for mounting on an automatic polishing machine.

GRINDING AND POLISHING

Procedures for grinding and polishing geological samples using several grades of diamond abrasive are described by Taggart (1977). Allen (1984) reported a one-stage precision polishing technique using aluminum oxide that gave fast, satisfactory results for a range of geological samples. Although polishing of very sensitive samples, such as some soft shales, is sometimes best carried out by hand, automatic grinding and polishing machines offer good results in the majority of cases (Fynn and Powell, 1979). If multiple grinding and polishing steps are employed, the initial grinding may be accomplished using a 600-grade grit. This process removes gross irregularities and any deformed surface material generated during cutting. A second stage of grinding is then normally employed using a finer grit (e.g., 6-μm diamond abrasive). The number of subsequent polishing stages selected should depend upon the nature of the samples and on the nature of the analyses to be performed on the samples. To obtain a very fine polish, some authors employ three polishing stages using different polishing cloths and 6-μm diamond paste, 1-μm diamond paste, and a very fine oxide polishing agent such as OP-S (e.g., Dijkshoorn and Fens, 1992). For many geological specimens, adequate results can be obtained using a single stage of polishing with 1-μm diamond abrasive.

Unless a sample contains soluble or water-sensitive components, grinding and polishing can be performed using water. With problematic samples, this operation can be performed using a mixture of abrasive in mineral oil, paraffin, or acetone.

STAINING

Although most materials show good compositional contrast in BSE images, it is sometimes desirable to make certain components more "visible" by staining. The need for greater contrast particularly applies to organic matter, which in some circumstances may be difficult to distinguish from epoxy resin or residual porosity. A number of staining procedures have been used for this purpose, probably the most common being exposure to vapors or dilute solutions of a heavy metal compound such as osmium tetroxide (e.g., Green et al., 1979). The osmium tetroxide is taken up preferentially by the organic matter in the specimen and is subsequently easily identified by its relatively high atomic number contrast.

SAMPLE COATING AND MOUNTING

The finished polished sections or chips are coated with 100–150 Å of carbon in a vacuum sputter coater prior to SEM examination. For most samples carbon is preferred for coating because gold gives rise to a high degree of scattering from the surface film, which tends to suppress the atomic number contrast from the underlying specimen. A carbon coat allows a high proportion of the incident electrons to pass through with minimum scattering while providing sufficient conductivity to prevent charging of the specimen.

Most SEM stages can be adapted to take sample holders designed for a range of specimen sizes and shapes, including standard polished thin sections or polished chips of similar size. If necessary, however, the specimen can be glued onto a 1-cm-diameter aluminum stub. In all instances it is good practice to place a trail of carbon dag or paint to ensure good electrical conductivity between the specimen and stub.

3

Sediment Grains, Weathering, and Early Diagenetic Phenomena

APPLICATION OF BSE TO STUDY OF SEDIMENT GRAINS

There are two main ways in which backscattered electron microscopy (BSE) images can be useful in the study of loose sediment grains. In both cases it is first necessary to embed a representative subsample of grains in a resin block to allow preparation of a polished surface (not necessarily a polished thin section). If it is necessary to examine the textural relationships of the grains *in situ*, without disturbance to the primary sediment fabric, resin impregnation can be carried out in the field. Otherwise, impregnation is normally undertaken in the laboratory after homogenization and splitting to provide a representative subsample.

As discussed in Chapter 2, the atomic number contrast evident in BSE images allows many of the common minerals in modern sediments and soils to be differentiated. In situations where there is a relatively small number of mineral species present in a sample, and where each species has a significantly different backscattering coefficient, it may be possible to determine quantitatively the relative abundance of different minerals. Figure 3.1 shows a heavy mineral beach placer deposit from Queensland, Australia, which consists of a mixture of quartz grains and heavy minerals, chiefly ilmenite and zircon. The abundance of heavy minerals in a series of such micrographs can be determined by grain counting, either manually or automatically by using dedicated computer software. By adjustment of the contrast mechanism, zircon and ilmenite can be differentiated and counted separately. The abundance of different minerals can be estimated by setting gray-level thresholds to define a "window" specific to each mineral detected (e.g., Ball and McCartney, 1981; Pye, 1984a). If a wide range of mineral types is present in a sample, including some with similar BSE coefficients and overlapping gray levels, it may be necessary also to employ automated X-ray microanalysis to differentiate and quantify the different mineral phases.

The second main way in which BSE images are useful in the study of individual grains relates to the quantification of grain size and form. Most previous studies of grain form in unconsolidated sediments have used "silhouette" images of grains that have been scattered on a light table or imaging plate, or photographed in the scanning electron microscope (SEM) using the secondary electron (SE) mode. The silhouetted grain outlines are then digitized and subjected to some form of computerized image analysis to quantify aspects of grain size and shape. Among the more widely used methods have been Fourier analysis (e.g., Ehrlich and Weinberg, 1970; Ehrlich et al., 1980) and fractal analysis (e.g., Orford and Whalley, 1983, 1991). Owing to the composite nature of the silhouette image, however, much of the finer textural detail,

Figure 3.1. A mixture of quartz and heavy mineral grains (ilmenite and zircon) in a beach placer deposit, Queensland, Australia. Scale bar = 100 μm.

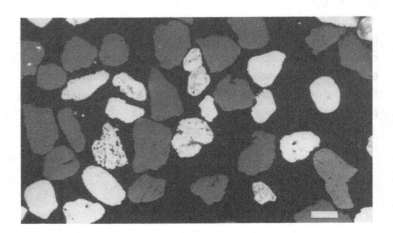

which in Fourier analysis is described by the higher-order harmonics, may be lost. In this instance, greater accuracy may be obtained by digitizing images obtained from a single-plane cut through grains that have been embedded in a resin block (e.g., Fig. 3.2). The nature of grain overgrowths and small-scale dissolution features is often much more visible when sectional images of grains are examined in this way. Although it is possible to digitize grain outlines from micrographs obtained by using the SE mode, the task is made much easier by using the BSE mode because better contrast can be obtained between the grains and the surrounding resin. Also, grains other than quartz can be more readily distinguished and excluded from the analysis. Figure 3.2 shows a number of subangular quartz grains collected from a podzolic weathering profile developed on late Quaternary coastal dune sands in northeastern Australia. The highly etched nature of the grains, and patchy development of secondary quartz overgrowths, is clearly evident. The intensity of postdepositional chemical weathering that has affected these grains is a reflection of the highly siliceous nature of the dune sands, the high rainfall in the area, and the abun-

Figure 3.2. Quartz grains from a weathered dune sand deposit at Ramsay Bay, North Queensland, Australia, embedded in a resin block. The highly etched nature of the grains and the presence of patchy quartz overgrowths are clearly evident. Scale bar = 100 μm.

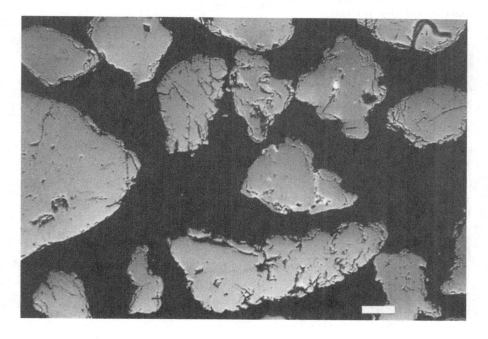

dance of humic acids released by the acidophyllous dune vegetation. Typical pH values of the podzolic soils in this area are less than 4 and in some instances fall below 2 (Pye and Mazzullo, 1994).

WEATHERING PHENOMENA

BSE images can be useful in studies that aim to quantify the relative degrees of weathering of different minerals in natural rock samples and building stones, or to identify the nature of weathering crusts and patinas developed on such materials. For example, Chalcraft and Pye (1984) used BSE to examine the nature of weathering on natural outcrops of Precambrian quartzite at Mount Roraima in southeastern Venezuela. They observed that breakdown of the rock proceeds primarily as a result of silica solution along quartz grain boundaries. Owing to solution, spaces between grains progressively widen and allow enhanced penetration of meteoric fluids into the rock, thereby enhancing the rate of weathering through a process of positive feedback. Any remaining feldspar in the rock is rapidly altered to kaolinite, some of which is redistributed along the grain-boundary cracks (Fig. 3.3).

Figure 3.3. Weathered Precambrian quartzite from Mount Roraima, Venezuela, showing widening of grain-boundary cracks between quartz grains as a result of silica dissolution, and infilling of cracks and pores by authigenic kaolinite. The white grain is anatase. Scale bar = 100 μm.

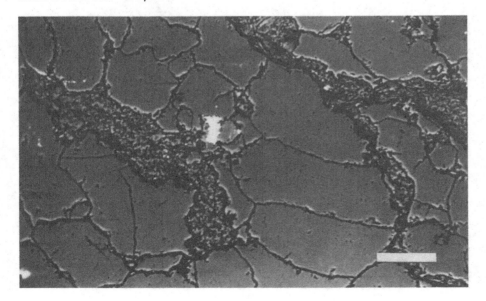

Pye (1985b) also demonstrated the importance of widening of grain-boundary cracks in the weathering of granitoid migmatites and gneisses in the semiarid Kora region of central Kenya. Relatively fresh rocks from this area possess very narrow grain-boundary cracks, and the constituent minerals show little evidence of internal physical or chemical alteration (e.g., Fig. 3.4). With progressive weathering, however, individual minerals take on an internally pitted appearance and the grain-boundary cracks become wider (Figs. 3.5 and 3.6). All minerals are affected to some degree, but feldspars and micas undergo greater alteration than does quartz. Secondary clay minerals, chiefly illite-smectite and kaolinite, together with traces of authigenic rutile or anatase, progressively replace parts of the micas, feldspars, and amphiboles, particularly close to grain-boundary cracks where rates of chemical alteration are highest (Fig. 3.7). The predominance of chemical alteration along grain boundaries eventually results in granular disintegration of the rock and produces a weathering regolith that has a high sand and silt content relative to clay. Not all generation of grain-boundary microcracks and crack-widening in rocks results from weathering. For example, Kanaori et al. (1991) described tectonically induced grain-boundary microcracks in granitic rocks around a fault zone in central Japan. Nonetheless, the existence of

Figure 3.4. Relatively unweathered granitoid gneiss from central Kenya. Q = quartz, KF = K-feldspar, M = mica, P = plagioclase, I = ilmenite. Grain-boundary cracks show no evidence of enlargement and grains show little internal pitting. Scale bar = 100 μm.

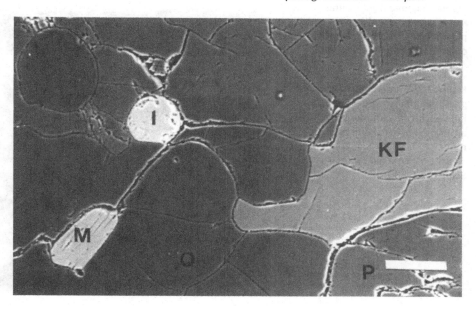

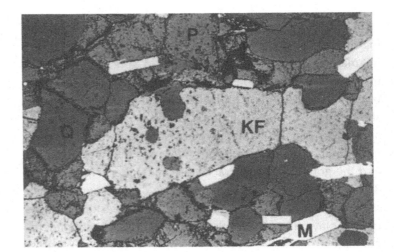

Figure 3.5. Moderately weathered granitoid gneiss from central Kenya. Q = quartz, KF = K-feldspar, P = plagioclase, M = mica. Widening of grain-boundary cracks and significant internal pitting are evident. Scale bar = 100 μm.

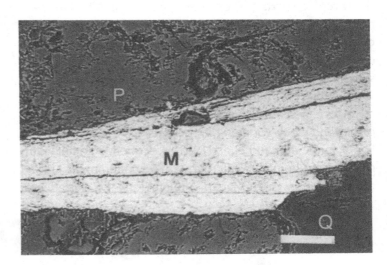

Figure 3.6. Moderately weathered granitoid gneiss from central Kenya, showing etching of plagioclase (P) and mica (M) and quartz (Q). Scale bar = 50 μm.

Figure 3.7. Altered biotite mica in granitoid gneiss from central Kenya. The development of authigenic clay (C) and anatase/rutile (white spots) can be seen clearly. M = mica, Q = quartz. Scale bar = 20 μm.

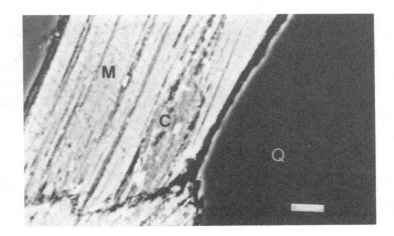

microcracks of whatever origin is likely to enhance the rate of weathering and rock-break disintegration.

Pye and Miller (1988) illustrated the value of BSE imaging in a study of shale weathering carried out in the context of a catastrophic failure of a shale embankment dam at Carsington, Derbyshire, U.K., in 1984. The study revealed that displacive growth of secondary gypsum crystals played an important role in disrupting the shale fabric and provided strong observational evidence for the importance of microbially mediated pyrite oxidation and dissolution of primary calcium carbonate minerals in the embankment fill. Oxidation of pyrite in the shale produced sulfuric acid, which leached much of the carbonate from the shale. The overall effect of these reactions was estimated to have increased the porosity of some parts of the shale fill by up to 10%, with possibly significant implications for its geotechnical behavior (Fig. 3.8).

Several studies have also employed BSE imaging in investigation of the weathering of building stone and other construction materials (e.g., Schiavon, 1992; Schiavon et al., 1994; Pye and Mottershead, 1995). Schiavon (1992) examined two texturally different oolitic limestones from urban areas in London and Cambridge, U.K., and showed that deterioration was primarily due to the crystallization of secondary gypsum

Figure 3.8. Chemically weathered shale from Carsington, Derbyshire, U.K. P = porosity formed by calcite dissolution; G = authigenic gypsum; FeS = pyrite. Scale bar = 20 μm.

both on the surface and within the stone. The limestone that had an oosparitic texture showed better resistance to decay than did the stone having an oomicritic texture, despite being exposed to higher pollution levels at its collection site in central London. Similar BSE examination of weathered patinas developed on granitic monuments in Spain and Portugal, combined with geochemical analytical techniques, allowed Schiavon et al. (1994) to demonstrate important differences in the composition and texture of the patinas. They concluded that these differences partly reflected differences in the intensity and nature of atmospheric pollution at the two sites.

Pye and Mottershead (1995) used BSE microscopy in an investigation of honeycomb weathering in a seawall at Weston-super-Mare, U.K. Carboniferous sandstone coping stones that cap the seawall display various degrees of honeycomb development related to varying exposure to marine salt spray. Underlying blocks of Carboniferous limestone that form the main body of the seawall do not show the development of honeycombs. BSE examination of polished sections of the sandstone showed that enlargement of the tafone takes place through the processes of microscale delamination and granular disintegration (Figs. 3.9 and 3.10). No *in situ* crystals of halite or other marine salt were observed, although the rocks were found to contain significant levels of chlorine that is apparently adsorbed onto other mineral phases, notably the clay matrix that is prominent in this particular sandstone (Pennant Sandstone from Gloucestershire). On the basis of textural, mineralogical, and geochemical evidence, it was concluded that the main weathering process is not a physical one involving the crystallization and expansion/contraction of salts, but rather a chemical-physical one in which ion-exchange processes between saline pore fluids (derived from sea spray) and clay minerals lead to expansion of the clay matrix and a gradual loss of intergranular cohesion.

BSE microscopy can also be used to good effect in investigations of the effects of weathering and chemical attack on concrete, render, and other manufactured construction materials. Table 3.1 lists the mean atomic number and backscattering coefficients of some compounds of ideal formula that are commonly present in cement and concrete, as either aggregate, cementation phases, or alteration products. Figure 3.11 illustrates secondary ettringite that has formed as a result of sulfate attack on a concrete floor slab. The formation

Figure 3.9. Microdelamination resulting in enlargement of tafoni in Pennant Sandstone, Weston-super-Mare, U.K. Scale bar = 200 μm.

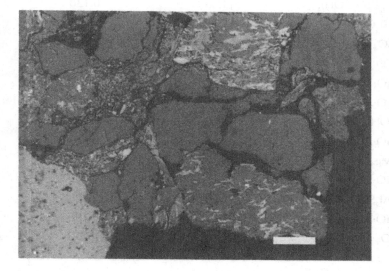

Figure 3.10. Granular disintegration during tafoni enlargement, Pennant Sandstone, Weston-super-Mare, U.K. Scale bar = 100 μm.

of large amounts of ettringite is detrimental to the strength of the concrete because this calcium-alumino-sulfate mineral contains 31 water molecules in its structure and is prone to large volume changes. These water molecules cause expansion of the concrete and progressively reduce its strength. In extreme cases, concrete that has been severely affected by sulfate attack becomes crumbly and disintegrates altogether. The small ettringite crystals shown in the BSE image in Figure 3.11 could not be distinguished by optical microscopy and were detected only in low amounts by X-ray diffraction analy-

TABLE 3.1

Mean atomic number ($\bar{Z}$) and calculated backscattering coefficients (η) for some minerals commonly found in cement and concrete

Mineral	Formula	Mean atomic number ($\bar{Z}$)	Backscattering coefficient (η)
Mg chlorite	$Mg_5Al_2Si_3O_{10}(OH)_4$	10.17	0.124
Kaolinite	$Al_4Si_4O_{10}(OH)_8$	10.24	0.125
Albite	$NaAlSi_3O_8$	10.71	0.132
Quartz	SiO_2	10.80	0.132
Ettringite	$3CaO \cdot Al_2O_3 \cdot 3CaSO_4 \cdot 31H_2O$	10.82	0.132
Dolomite	$CaMg(CO_3)_2$	10.87	0.133
Muscovite	$K_2Al_6Si_6O_{20}(OH)_4$	11.33	0.138
Orthoclase	$KAlSi_3O_8$	11.85	0.145
Gypsum	$CaSO_4 \cdot 2H_2O$	12.12	0.146
Calcite	$CaCO_3$	12.57	0.150
Calcium silicate hydrate	$3CaO \cdot 2SiO_2 \cdot 3H_2O$	13.07	0.157
Dicalcium silicate	$2CaO \cdot SiO_2$	14.00	0.169
Portlandite	$Ca(OH)_2$	14.33	0.171
Tricalcium aluminate	$3CaO \cdot Al_2O_3$	14.33	0.173
Biotite	$K_2(Mg_2Fe_4)Al_2Si_6O_{20}(OH)_4$	14.59	0.174
Tricalcium silicate	$3CaO \cdot SiO_2$	15.06	0.182
Fechlorite	$Fe_5Al_2Si_3O_{10}(OH)_4$	16.05	0.188
Tetracalcium aluminoferrite	$4CaO \cdot Al_2O_3 \cdot Fe_2O_3$	16.65	0.198
Hematite	Fe_2O_3	20.59	0.238

sis. Nevertheless, their development has had a serious effect on the bond strength between the aggregate particles and the hardened cement paste. In this instance the sulfate attack resulted from the upward movement of sulphate-bearing groundwater that originated from a layer of industrial ash fill beneath the floor slab.

EARLY DIAGENETIC PHENOMENA

A clear distinction between weathering and early diagenesis is sometimes difficult to draw, because both processes commonly involve dissolution and alteration of primary (detrital) mineral grains and the formation of new, secondary phases. The latter may form grain replacements, coatings, or intergranular cements. Figure 3.12 shows a sample of marine mud that contains a high proportion of angular volcanic glass shards. Such glassy materials are highly susceptible to dissolution and chemical alteration commonly associated with the formation of authigenic clay minerals (especially smec-

Figure 3.11. Development of secondary ettringite (E) in a concrete floor slab affected by sulfate attack. Scale bar = 50 μm.

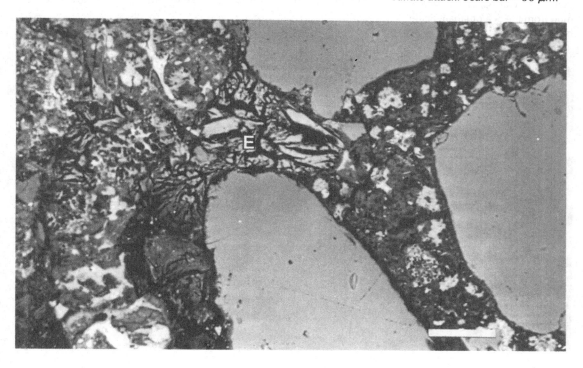

tites) and zeolites. Feldspars and other detrital crystalline minerals may undergo similar rapid alteration, particularly where hydrothermal fluids permeate the sediment or in saline and alkaline lake environments. Figure 3.13 shows a partially dissolved feldspar grain in a modern mud sample from Lake Magadi, an alkaline carbonate lake in southern Kenya. Alteration of feldspars and volcanic glass in these sediments is accompanied by widespread formation of authigenic zeolites, clay minerals, and unusual silicate minerals such as magadiite, kenyaite, and makatite.

In areas of the deep sea floor where sedimentation rates are relatively low, and where geochemical gradients in the underlying sediments are quite sharp, diagenetic phenomena such as formation of manganese nodules may occur. The nodules may range in size from a few tens of micrometers to several centimeters. Some are relatively pure, consisting largely of one or more types of manganese oxyhydroxide; others contain detrital impurities and significant amounts of other authigenic phases (e.g., iron oxyhydroxides, iron-rich silicate minerals, or iron-manganese carbonates). In BSE images, slight variations in chemical composition are clearly revealed as the concentric growth patterns that are charac-

Figure 3.12. Abundant volcanic glass shards (VG) and pelloid of authigenic clay (C) in a marine mud from the Sea of Japan. Scale bar = 100 μm.

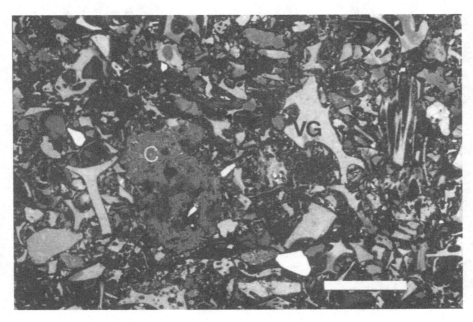

teristic of many manganese nodules (Fig. 3.14). In detail, the compositional zones commonly show an undulatory or mamillary pattern that reflects the growth pattern of the nodule (Fig. 3.15).

Diagenetic nodules and concretions are also common features in terrestrial soils and sediments. In some instances, the concretions are cemented by a passive cement that partially or wholly infills the primary (and some secondary) pore spaces. For example, Pye (1984b) used BSE microscopy to examine siderite-Mg-calcite-iron monosulfide concretions that are actively forming in intertidal sandflat and saltmarsh sediments near Warham, north Norfolk, U.K. In the sandier facies, siderite and Mg-calcite initially form grain coatings that progressively grow and coalesce to infill much of the original pore space, as shown in Figure 3.16a. Figures 3.16b and 3.16c are images of the same area in which gray-level "windows" have been set to show only the distribution of

Figure 3.13. Partially dissolved detrital volcanic feldspar in a matrix of fine-grained mud (partly authigenic), Lake Magadi, Kenya. Scale bar = 20 μm.

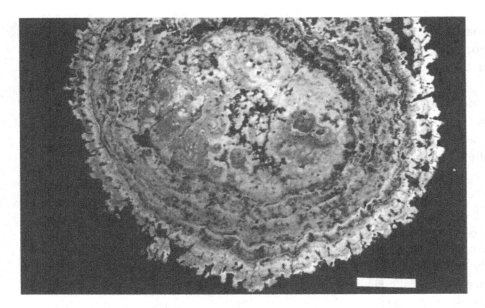

Figure 3.14. A manganese nodule from the Pacific Ocean, showing concentric growth banding. Scale bar = 1000 μm.

Figure 3.15. Detail of internal compositional zonation within a manganese nodule from the Pacific Ocean. Scale bar = 100 μm.

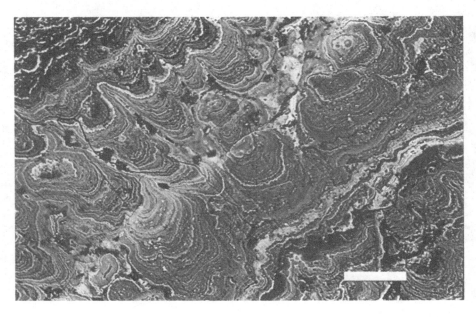

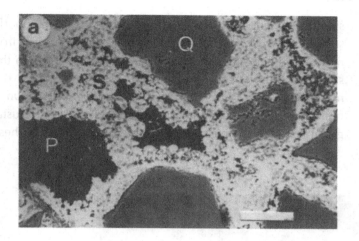

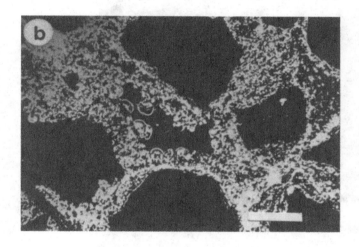

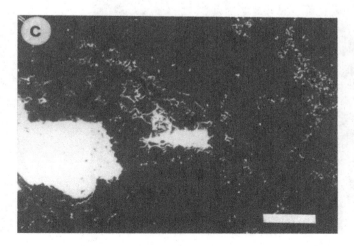

Figure 3.16. (a) BSE image of a siderite-cemented concretion from Warham, North Norfolk, U.K. (Q = quartz; P = porosity; S = siderite); (b) windowed BSE image showing distribution of siderite cement (white); (c) windowed and inverted BSE contrast image showing distribution of porosity (white). Scale bars = 100 μm.

Figure 3.17. Authigenic siderite (white = S) replacing a gastropod shell, Mg-calcite (C) lining internal chamber void, in concretion from Warham, Norfolk, U.K. Scale bar = 200 μm.

Figure 3.18. Higher-magnification view of part of Figure 3.17, showing coating of iron monosulfide (FeS, arrowed) on Mg-calcite crystals. Scale bar = 100 μm.

siderite cement and unfilled porosity. In the more poorly sorted muddy marsh facies, siderite, Mg-calcite, and iron monosulfide crystals are typically much smaller than in the sandy facies and occur as a partially displacive cement. The morphology of the cement crystals is dependent upon a number of factors, including the available space for crustal growth and the chemistry of the ambient pore fluids. In these

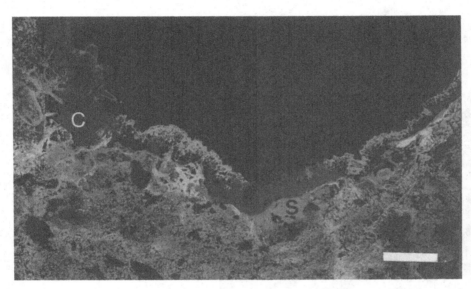

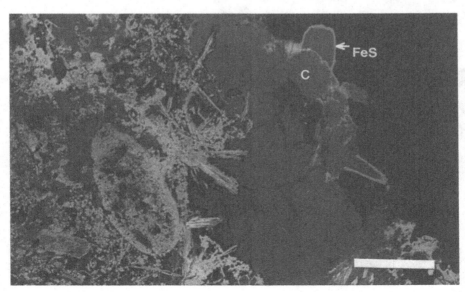

concretions, siderite occurs in a number of different forms including aggregates of intergrown rhombic crystals, botryoidal and mamillary aggregates of radial-fibrous crystals, and irregular clusters of acicular crystals. Mg-calcite occurs mainly as equant, blocky crystals. The different crystal forms of the same mineral commonly occur in close proximity, or are intergrown, as shown in Figures 3.17–3.22. Quantitative

Figure 3.19. Acicular and radial-fibrous siderite lining an internal void in a concretion from Warham, U.K. Scale bar = 100 μm.

Figure 3.20. Higher-magnification view of the radial-fibrous siderite shown in Figure 3.19. Scale bar = 20 μm.

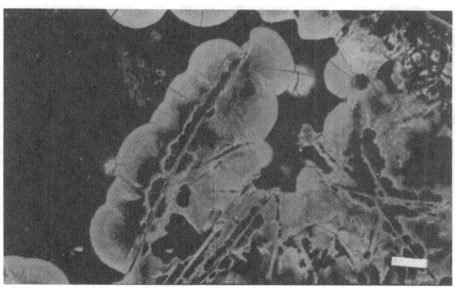

41

Figure 3.21. Acicular siderite lining an internal void in a concretion from Warham, U.K. Scale bar = 100 μm.

Figure 3.22. Higher-magnification view of part of Figure 3.21, showing two-stage siderite coating on quartz grains. Scale bar = 20 μm.

energy-dispersive analysis has determined that the Warham siderite contains little substituted Mg (< 3%), although Ca substitution may be up to 10 mol % (Fig. 3.23). Many of the concretions contain well-preserved root material in which the cells are both infilled and partially replaced by authigenic siderite and iron monosulfide (Allison and Pye, 1994;

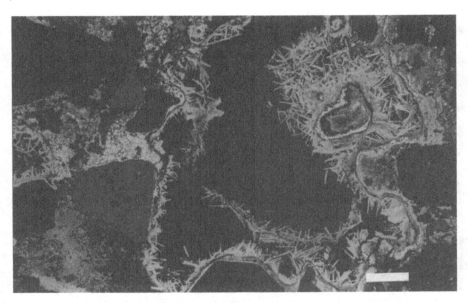

Figs. 3.24 and 3.25). Many shells within the concretions are partially or wholly replaced by siderite (Figs. 3.26 and 3.27), detrital aragonite being more susceptible to replacement than detrital calcite. Field experimental studies have shown that the process of shell replacement and concretion formation is extremely rapid, taking on the order of 10–50 years. Stable isotope and other geochemical data strongly suggest that microbial sulfate reduction is an important process involved in the formation of concretions (Pye et al., 1990).

Cementation and replacement by secondary carbonate minerals is also an important process in many arid and semi-arid zone soils. Pedogenic calcareous horizons, cemented mainly by calcite, dolomite, or a mixture of the two, may become exposed by surface erosion, leading to the formation of duricrusts. Following exposure, a sequence of partial dissolution and reprecipitation typically ensues, producing a

Figure 3.23. Ionic composition of siderite cement in concretions from Warham, U.K., determined by electron microprobe analysis.

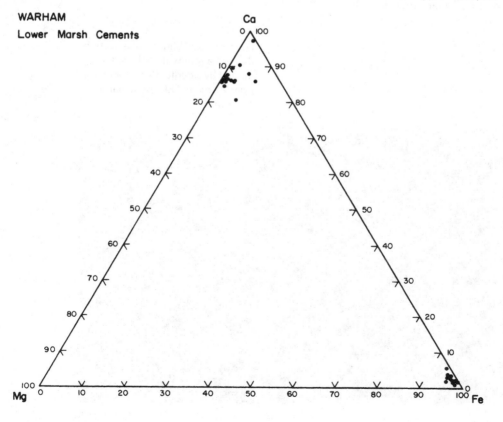

43

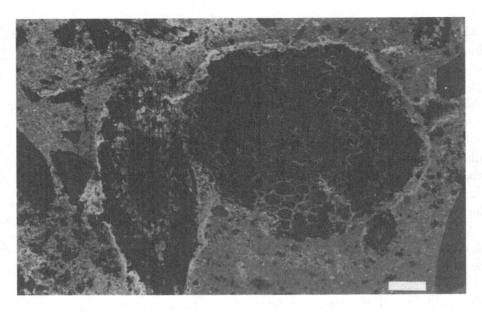

Figure 3.24. Cross-section through fossil roots, cell walls lined and replaced by siderite, in a concretion from Warham, U.K. Scale bar = 100 μm.

Figure 3.25. Oblique cross-section through a root with cells partly infilled by siderite, in a concretion from Warham, U.K. Scale bar = 100 μm.

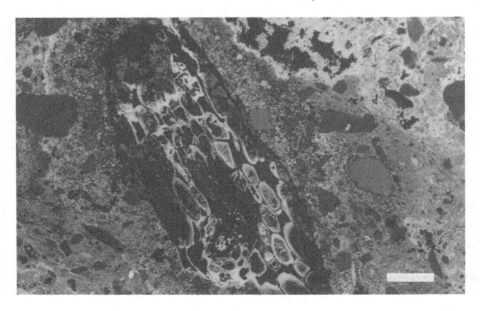

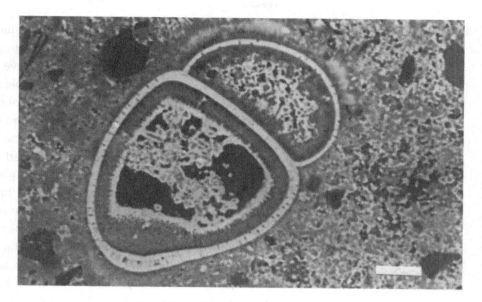

Figure 3.26. Small *Hydrobia* shell replaced and partly infilled by siderite (white), in a concretion from Warham, U.K. Scale bar = 20 μm.

Figure 3.27. Part of an aragonitic shell largely replaced by siderite (S), in a concretion from Warham, U.K. Remaining areas of aragonite (A) appear relatively dark. Scale bar = 100 μm.

complex textural and mineralogical compositional pattern in the rock. Figure 3.28 shows a relatively young groundwater calcrete collected from a dry stream bed developed on granitoid gneiss in semiarid Kenya (Pye and Coleman, in press). The rock has a high detrital matrix to cement ratio, the detrital component consisting mainly of sand- and silt-size quartz and feldspar fragments, with subsidiary mica. The cement is composed of low-Mg calcite that is dispersed within the matrix. By contrast, the BSE micrograph in Figure 3.29 shows part of a much older dolocalcrete that has formed within a soil profile developed on biotite-hornblende schist. This rock has a low ratio of detrital material to "cement," reflecting a much longer history of grain replacement and carbonate dissolution/reprecipitation. Much of the rock is composed of authigenic dolomite crystals, reflecting the high availability of Mg ions from the base-rich underlying bedrock. The dolomite areas contain a few residual detrital grains, which are clearly in the process of dissolution and replacement (Fig. 3.30). Manganese oxide derived from the alteration of micas and amphiboles forms discrete patches and dendrites within the rock (Fig. 3.31). Locally, authigenic Mg-rich clays (mainly palygorskite and smectite) are present as discrete pellets or more rarely as void linings and infillings (Fig. 3.32). The near-surface layers of the dolocretes are fractured and sometimes capped by a boulder zone, indicating partial breakup of the surface layer owing to weathering and dis-

Figure 3.28. Calcite cement (C) in the matrix of a late Holocene groundwater calcrete, from central Kenya. Scale bar = 50 μm.

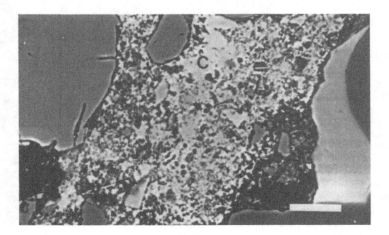

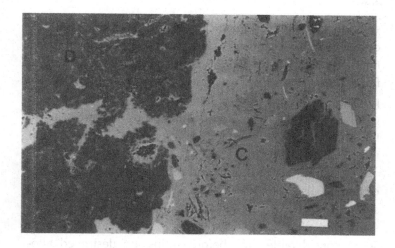

Figure 3.29. Dolomite (D) and calcite (C) in a Quaternary dolo-calcrete, from central Kenya. Note the greater abundance of silici-clastic inclusions in the calcite-filled fissure. Scale bar = 10 μm.

Figure 3.30. Partially dissolved amphibole grain in Quaternary dolocrete, from central Kenya. Scale bar = 30 μm.

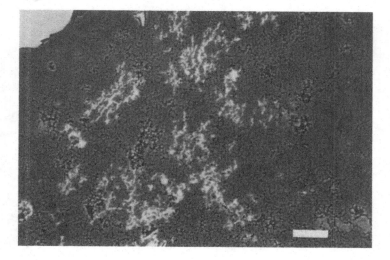

Figure 3.31. Manganese oxide veining (white) in Quaternary dolocrete, from central Kenya. Scale bar = 30 μm.

solution following stripping of the overlying sediments. The dolomite in these zones is cross-cut by veins in which the main cement is low-Mg, nonferroan calcite, surrounding fairly numerous siliciclastic inclusions. Dissolution of the primary dolomite in the upper part of the profile has evidently been followed by reprecipitation of the carbonate in the form of calcite, while part of the Mg has been incorporated into authigenic clays.

Detrital grain dissolution and replacement, either partial or total, is also common in evaporite deposits. Etching of the surfaces of quartz and feldspar grains is relatively rapid in saline and alkaline environments, although it normally takes some considerable time before grains are destroyed altogether. On the other hand, detrital fine-grained particles, such as clays, may be replaced quite quickly. Figures 3.33 and 3.34 show a number of angular, etched quartz and feldspar grains included in a matrix of authigenic gypsum. Formation of the gypsum crystals took place mainly displacively, close to the water table within eolian sand host sediments. Host grains that became trapped within the growing nodular layer subsequently experienced partial dissolution.

Authigenic precipitates, containing various amounts of included detrital material, have been widely reported from subglacial settings. A variety of cement types has been recorded, including calcite, amorphous silica, iron oxyhydrox-

Figure 3.32. Pelloid of authigenic palygorskite (P) in Quaternary dolocalcrete, from central Kenya. The matrix consists mainly of a mixture of calcite (C) and dolomite (D). Remnants of potassium feldspar (KF) and corroded quartz (Q) are also present. Scale bar = 30 μm.

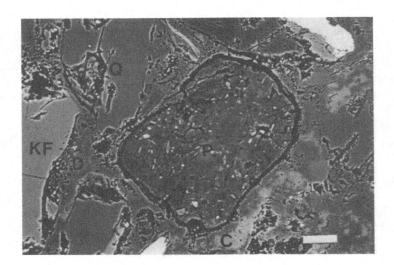

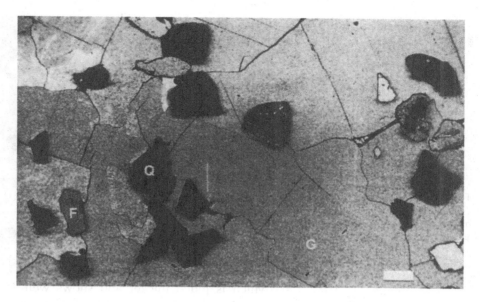

Figure 3.33. Detrital quartz (Q) and feldspar grains (F) contained within a gypsum nodule (G), from Libya. Scale bar = 100 μm.

ides, manganese oxyhydroxides, and jarosite (iron potassium aluminum sulfate hydrate). The formation and composition of these precipitates can be important in terms of understanding the nature of ice–bedrock interactions, subglacial water chemistry, and the dissolved load fluxes within glaciated catchments. Figure 3.35 illustrates the texture of a predominantly calcitic subglacial precipitate from Tsanfleuron, Switzerland, which has a two-layer structure. The lower layer,

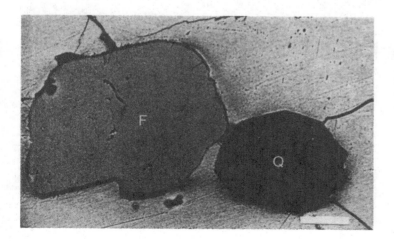

Figure 3.34. Higher-magnification view of two of the grains from the same nodule shown in Figure 3.33. F = feldspar and Q = quartz. Scale bar = 50 μm.

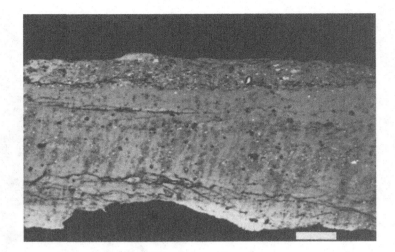

Figure 3.35. Laminated subglacial calcite precipitate from Tsanfleuron, Switzerland. Scale bar = 300 μm.

which is much thicker and shows well-developed small-scale lamination, consists of more than 85% calcite, whereas the thin and more discontinuous upper layer contains less than 50% calcite, the remainder comprising detrital siliciclastic grains. Figure 3.36 shows a contrasting goethite/jarosite subglacial precipitate from Mitdalsbreen, Norway. In this instance the goethite and jarosite have penetrated fissures in the bedrock underlying the surficial precipitate. The composition of the included grains in another precipitate from Mitdalsbreen shown in Figure 3.37 is different from that of the underlying schist, indicating that the grains are predominantly of allochthonous origin.

Figure 3.36. Geothite-jarosite subglacial precipitate (white) from Mitdalsbreen, Norway. Goethite and jarosite penetrate along fissures in the underlying bedrock. Scale bar = 100 μm.

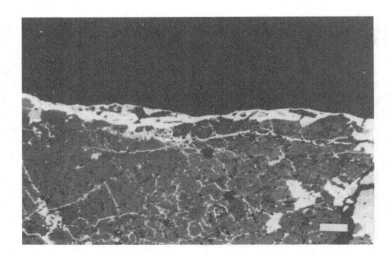

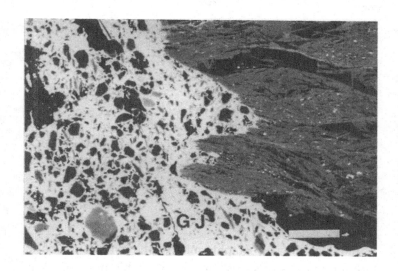

Figure 3.37. Goethite-jarosite (G-J) subglacial precipitate (white) developed on schist bedrock, from Mitdalsbreen, Norway. Many of the grains included within the precipitate are silt- to sand-size quartz and feldspars, suggesting that a significant proportion of the grains has been derived from an allochthonous source. Scale bar = 100 μm.

4

Sandstones

Although the petrographic microscope remains the basic tool for analysis of sandstones, petrographic microscopy is being increasingly supplemented by newer analytical methods. Backscattered electron microscopy (BSE) ranks as one of the more useful of these newer techniques because BSE, coupled with energy-dispersive X-ray analysis (EDX), provides a powerful tool for mineral identification, porosity determination, and study of fine-scale grain-to-grain and intragrain relationships (Chapter 3). BSE and EDX can be used, for example, to differentiate easily among such mineral phases as quartz, feldspars, clay minerals, heavy minerals, pyrite, and carbonate and silica cements in sandstones. Mineral identification using BSE and EDX can be particularly advantageous in studying very fine-grained sandstones that are difficult to analyze with a petrographic microscope. Furthermore, techniques exist that allow quantitative mineral identification by automated image analysis techniques – a potential saving in operator time. BSE also allows identification of open pores as small as 0.1 μm; therefore, it is an extremely useful tool for studying porosity and microporosity. Finally, BSE provides a superior method for differentiating within single mineral grains different mineral phases that may have originated through diagenetic processes (e.g., replacement, albitization, dissolution, and precipitation).

QUANTITATIVE IDENTIFICATION OF
FRAMEWORK CONSTITUENTS

The fundamental techniques for identifying minerals by using BSE are discussed in Chapters 2 and 9. In these chapters, we describe quantitative techniques for identifying different mineral phases through a combination of X-ray mapping and BSE image analysis, which utilize differences in gray-scale brightness of different mineral phases as displayed in BSE (Tovey and Krinsley, 1991). The BSE method is particularly important in study of shales because of the difficulty of applying petrographic microscopic methods to identification of very fine minerals in shales (Chapter 5).

Similar BSE methods can be applied to identification of minerals in sandstones (e.g., Dilks and Graham, 1985). As discussed in Chapter 2, the intensity or gray level of a given mineral phase in a BSE image is determined by the backscatter coefficient, η, which is a function of the atomic number of the elements that make up the mineral. The backscatter coefficients of some common minerals in sandstones are given in Table 2.2; the higher the backscatter coefficient, the brighter the mineral appears in BSE. An example of this difference in brightness is shown in Figure 3.1, which shows the relative intensities of quartz and heavy minerals as they appear in BSE. Initial identification of minerals is made by using EDX to determine the relative abundance of chemical elements in the mineral. Once a given mineral has been identified, its brightness or intensity is determined on the basis of a BSE image gray-level scale extending from 0–255. Segmentation thresholds are then set to define windows for each mineral phase in the gray-level spectrum (e.g., 56–97 for kaolinite, 98–109 for quartz, 110–131 for K-feldspar). Once the segmentation thresholds for each mineral are set, the image analyzer can automatically determine the area of each mineral in a given image field and thus can calculate the relative abundance of that mineral in the image field

(Dilks and Graham, 1985). By picking a number of image fields (say, 1.0 × 1.0 mm in size) throughout a given specimen, the average abundance of each mineral in the entire specimen can be determined. Not all minerals may be discriminated on the basis of gray level, because the gray levels of some minerals (e.g., quartz and feldspar) may be too similar. In such cases, additional methods, such as X-ray mapping of chemical composition, may be required for identification, as discussed in Chapter 9.

Using BSE to identify framework minerals in sandstones has some advantages and some severe limitations compared to using petrographic microscopy. Advantages include the fact that some minerals can be identified more accurately by a combination of BSE and EDX than by petrographic microscopy (e.g., differentiating quartz and untwinned plagioclase feldspar). Resolution in BSE is much better than that in optical microscopy; therefore framework grains of very fine size (fine sand and silt) are much easier to study using BSE techniques. Finally, the potential exists to automate the identification process, as described above, so that identification can proceed with little oversight from the operator, yielding considerable savings in operator time (Chapter 9). A major limitation of the BSE method is that BSE cannot itself distinguish between images of different minerals having the same intensity (gray-scale brightness). A particular problem arises in identification of rock fragments, which may be composed of many different kinds of fine-grained minerals. No published studies have focused on the problem of identifying, and distinguishing among, rock fragments by utilizing BSE. Because rock fragments are prevalent in many sandstones (lithic arenites), difficulty in distinguishing rock fragments may prove to be a major limitation in automated mineralogical identification of lithic sandstones.

MATRIX MINERALS AND CEMENTS

Most sandstones contain various amounts of matrix minerals and/or cements. Owing to the low resolution of optical microscopes, petrographic identification of fine-grained matrix minerals and cements and study of their fine-scale textural relations is virtually impossible. On the other hand, the resolution of backscattered electron detectors is such that the textural details of pore-filling matrix minerals and cements are intimately revealed. Furthermore, the use of EDX permits detailed quantitative geochemical and mineralogical study of the pore fillings. Such detailed study may, among other things, allow an investigator to determine if matrix minerals (clay minerals, fine-grained quartz and feldspars) are detrital or authigenic. For example, Huggett (1984) studied hydromuscovite and kaolinite intergrowths in the pores of Coal Measure sandstones in the United Kingdom. By using BSE imaging in combination with wavelength-dispersive spectral analysis, she was able to determine that the muscovite grains were detrital, had undergone incipient alteration to illite, and occur intergrown with chlorite and kaolinite; the kaolinite is authigenic and occurs as pore-filling books.

Figure 4.1 provides an additional example of this kind. In this example, detrital biotite in a Carboniferous sandstone from the North Sea has been split by growth of illite and kaolinite between the cleavage planes. Figure 4.2 shows fibrous illite cement partially lining a pore in a Rotliegend sandstone from the southern North Sea, and Figure 4.3 shows pore space in a Carboniferous sandstone from the southern North Sea that is almost completely infilled by diagenetic illite.

In addition to clay minerals, many other kinds of cements in sandstone are usefully studied by BSE image analysis. For example, Purvis (1989) used BSE and electron probe microanalysis to identify and study magnesite cement in Rotlie-

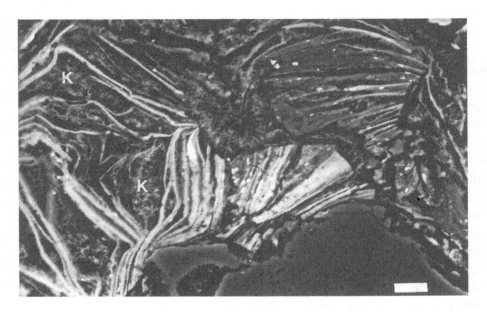

Figure 4.1 (above). Detrital biotite grain split by growth of kaolinite (K) between cleavage planes. Carboniferous, southern North Sea. Scale bar = 20 μm.

Figure 4.2 (below). Sandstone pore lined with fibrous illite cement (fine crystals). Rotliegend sandstone, southern North Sea. Cementation by illite significantly reduces the porosity of these reservoir sandstones. Scale bar = 200 μm.

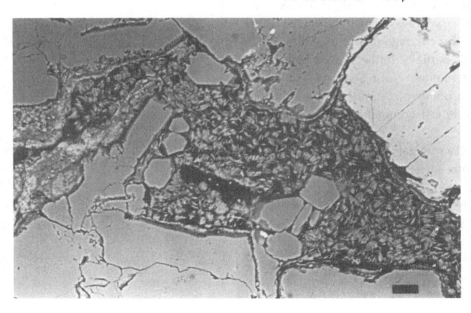

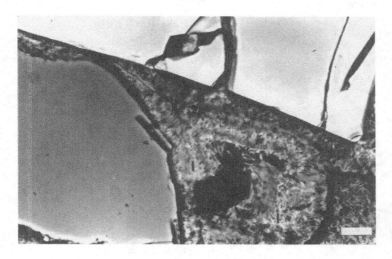

Figure 4.3. Sandstone pore almost infilled by diagenetic illite and minor kaolinite. The illite cement (I) fills both a primary pore and secondary pore space formed by dissolution of a detrital framework grain. Carboniferous, southern North Sea. Scale bar = 100 μm.

gend sandstones of the southern North Sea, and Pye and Krinsley (1986a) studied carbonate and evaporite cements in eolian Rotliegend sandstones from the same area (Figs. 4.4, 4.5, and 4.6). Pye (1984b) applied the BSE technique to the study of siderite cements in marsh sediments of Norfolk, England. Figure 4.7 shows hematite cementing quartz grains in Quaternary dune sand, North Queensland, Australia.

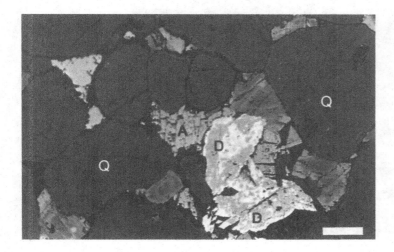

Figure 4.4. Authigenic, zoned, ferroan dolomite (D); anhydrite (A); and quartz (Q) infilling a pore in Rotliegend sandstone, southern North Sea. Scale bar = 100 μm.

Figure 4.5 (above). Complex
carbonate and kaolinite cement
filling sandstone pore space,
Rotliegend sandstone, southern
North Sea. D = dolomite with
lighter zones of ferroan dolomite,
K = kaolinite. Scale bar = 20 μm.

Figure 4.6 (below). Zoned dolomite
rhomb (center of photograph) in illite-
lined pore. The dark core of the rhomb is
composed of dolomite, with lighter bands
of ferroan dolomite. The lighter outer
zone is ankerite, which appears mottled
owing to variations in Fe/Mg ratio. The
dark grains are quartz. Rotliegend
sandstone, southern North Sea. Scale
bar = 50 μm.

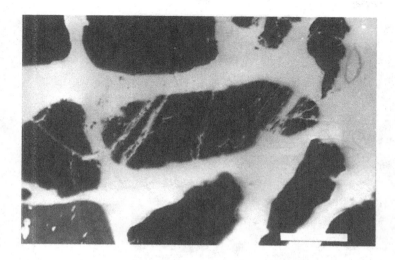

Figure 4.7. Quartz grains (dark) cemented by hematite (bright). Quaternary dune sandstone, North Queensland. Scale bar = 100 μm.

OTHER DIAGENETIC PHASES

One of the most useful applications of BSE to the analysis of sandstone is in the study of the fine-scale relations between and among authigenic mineral phases that form as a result of replacement reactions. For example, framework grains of quartz, feldspars, and rock fragments may be partially or entirely replaced by a variety of minerals, including microquartz (chert), calcite, dolomite, siderite, anhydrite, barite, clay minerals, zeolites, and hematite. Also, matrix minerals and cements may be replaced by other minerals. BSE allows examination of replacement textures on a scale ranging from millimeters to microns. Therefore, it is possible to recognize different mineral phases within a single grain and work out the paragenetic sequence of replacement. The relationship between replacement minerals and characteristics of the host mineral such as fractures, micropores, and cleavage planes is revealed with unusual clarity, to a degree that cannot be approached by optical microscopy.

Figures 4.8–4.11 provide several examples of replacement textures in sandstones. Figure 4.8 shows a detrital feld-

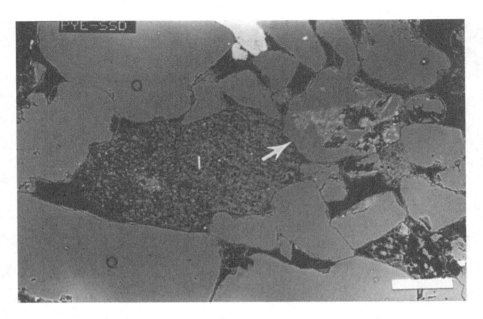

Figure 4.8 (above). Detrital feldspar grain (center) completely replaced by authigenic illite (I). Note partial dissolution of detrital quartz grain (arrow). Q = quartz. Rotliegend Sandstone, southern North Sea. Scale bar = 100 μm.

Figure 4.9 (below). Detrital K-feldspar grain partially dissolved and replaced by chlorite, Rotliegend sandstone, southern North Sea. 1 = chlorite, 2 = kaolinite, 3 and 4 = illite, 5 = K-feldspar; large gray grains are quartz. Scale bar = 100 μm.

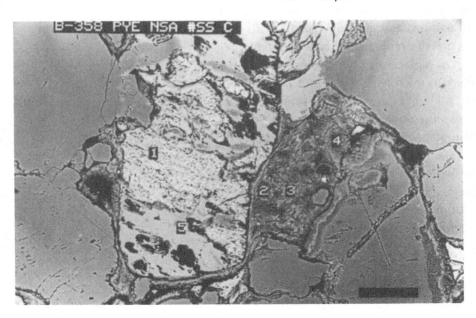

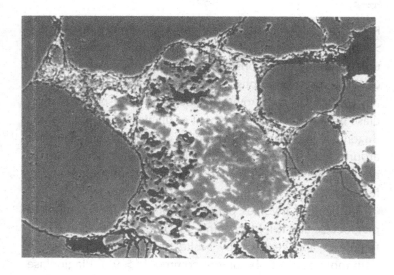

Figure 4.10. K-feldspar grain (center of photograph) selectively dissolved (black areas) and replaced by quartz (dark gray), Torridonian Sandstone (Precambrian), Scotland. Scale bar = 100 μm.

spar grain that has been replaced entirely by illite, and Figure 4.9 shows partial dissolution of K-feldspar and partial replacement by chlorite. In Figure 4.10, a K-feldspar grain has been partially dissolved and partially replaced by quartz. Barite in Figure 4.11 infills secondary pores as a cement and also replaces authigenic illite.

Two special kinds of replacement processes involving feldspars are replacement of plagioclase feldspar or detrital K-feldspar by authigenic K-feldspar and replacement of placioclase and K-feldspar by albite (albitization). Authigenic

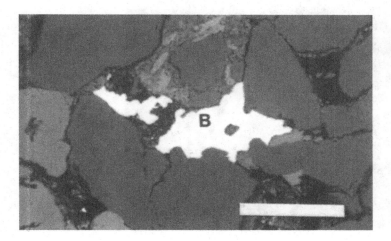

Figure 4.11. Barite (B) infilling secondary pore space and replacing authigenic illite. Rotliegend sandstone, southern North Sea. Scale bar = 100 μm.

K-feldspar is commonly chemically pure (mole percent Or > 98), is untwinned, and may be heavily clouded by vacuoles and tiny inclusions or be free of such inclusions. It can occur as pseudomorphs after detrital K-feldspar (Morad et al., 1989) or as overgrowths on detrital K-feldspar or plagioclase grains. Partial replacement of plagioclase grains by authigenic K-feldspar may also occur along cleavage planes or around fractures or micropores (Fig. 4.12). It is commonly not possible to identify such small zones of authigenic K-feldspar with a conventional petrographic microscope.

Albitization involves partial to complete replacement of plagioclase and K-feldspar (and, in some cases, volcanic rock fragments and zeolites) by albite (sodium-rich plagioclase with mole percent Ab content commonly > 98). Plagioclase or K-feldspar that has been completely replaced by albite has a uniform gray scale in BSE images but can be identified easily by using EDX owing to its high sodium content. Partially albitized feldspars display a patchy pattern of gray-scale intensities, with albitized (sodium-rich) areas appearing much darker than the nonalbitized parts of detrital plagioclase or K-feldspar grains (Figs. 4.13–4.15). Very com-

Figure 4.12. Authigenic K-feldspar (Kf, bright) partially replacing plagioclase (Pg, gray) along fractures. Miocene sandstone, Ocean Drilling Program Leg 127, Site 796, Japan Sea. Scale bar = 50 μm.

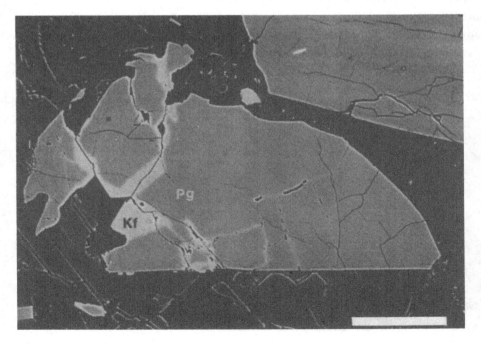

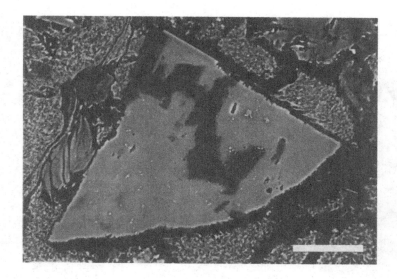

Figure 4.13. Partial, patchy replacement of a K-feldspar grain by albite (dark). Miocene sandstone, Ocean Drilling Program Leg 127, Site 797, Japan Sea. Scale bar = 25 μm.

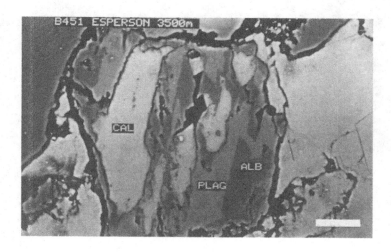

Figure 4.14. Calcic plagioclase feldspar (PLAG) partially replaced by albite (ALB) and calcite (CAL), Tertiary sandstone, Texas Gulf Coast, depth 3500 m. Scale bar = 20 μm.

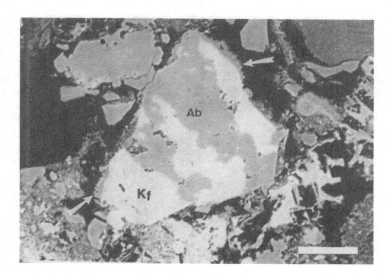

Figure 4.15. Albite cement forming overgrowths (arrows) on a partially albitized (Ab) K-feldspar grain (Kf). Miocene sandstone, Ocean Drilling Program Leg 127, Site 797, Japan Sea. Scale bar = 25 μm.

Figure 4.16 (above). A plagioclase feldspar grain (Pg) selectively altered to albite (Ab) around inclusions of glass (G). Miocene sandstone, Japan Sea. Core 127-797-25R, ~700 m below sea floor. Scale bar = 25 μm.

Figure 4.17 (right). Authigenic blocky albite (dark) selectively replacing detrital K-feldspar (light). Note relation of albite to large dissolution pores (black). Miocene sandstone, Japan Sea. Core 128-799B-61R, ~1020 m below sea floor. Scale bar = 50 μm.

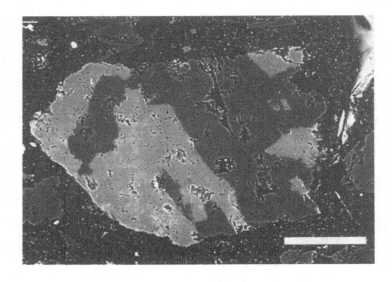

64

monly, patchy albitization is associated with secondary pores, fractures, and cleavages. Study of albitization textures in sandstones utilizing BSE and EDX is becoming increasingly common (e.g., Boggs and Seyedolali, 1992; Milliken 1989; Morad et al., 1990; Saigal et al., 1988; Seyedolali and Boggs, 1996). Several kinds of albitization textures can be recognized in BSE images, including: (1) precipitates of albite as ordinary cements in spaces not formerly occupied by detrital grains (e.g., as overgrowths on detrital grains; Fig. 4.15), (2) vein-like albite that partially replaces feldspar grains around glass inclusions (Fig. 4.16) or along crystallographic planes (see Fig. 4.22), and (3) blocky albite that may develop in both plagioclase and K-feldspar (Fig. 4.17) and that is commonly associated with secondary pores.

POROSITY AND MICROPOROSITY DETERMINATION

Evaluation of the size, shape, continuity, and total volume of pores in sandstones is of significant interest to geologists in terms of evaluating the diagenetic history of the sandstone and its reservoir characteristics (ability to transmit and store groundwater or hydrocarbons). Pore space in a sandstone can be studied with a petrographic microscope if the pores are impregnated with a dyed epoxy (commonly blue) prior to grinding thin sections. Petrographic analysis of porosity is, however, very time consuming and does not work well with very fine pores. BSE image analysis provides a much better method of studying porosity, particularly microporosity. The atomic number contrast between the epoxy resins used to impregnate and prepare polished samples for BSE analysis and the minerals in a specimen is quite large. Thus, epoxy appears much darker than most minerals, allowing epoxy-filled pore space to be readily distinguished from mineral grains (Fig. 4.18).

The size, shape, and continuity of pores can be easily

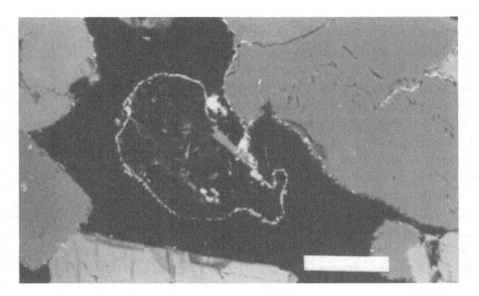

Figure 4.18. Large pore (black) among quartz grains (light gray) in a sandstone. The area outlined in white in the middle of the pore is a machine artifact. Scale bar = 50 μm.

studied by BSE, including micropores (e.g., within matrix clay minerals) that are much too small to be seen with a petrographic microscope (Nadeau and Hurst, 1991; Pye, 1984a). Furthermore, automated techniques can be used to determine quickly the total pore space (porosity) in a sample. Quantification of porosity is accomplished by setting the gray-scale window to exclude all the darker gray tones (epoxy) until only mineral surfaces remain. The percentage of gray tones deleted is equal to the porosity of the area . By examining several image areas in a given specimen and averaging the porosity in these areas, a representative average value for porosity of the entire sample can be determined.

Using BSE to study porosity is a particularly effective technique for identifying and studying the detailed shapes of secondary pores that originate by diagenetic processes owing to fracturing, shrinkage, or dissolution of framework grains, matrix, or cement (Pye and Krinsley, 1985). For example, Figure 4.19 shows microporosity in quartz grains that developed as a result of incipient dissolution of the quartz, and Figure 4.20 shows microporosity within a K-feldspar grain that resulted also from partial dissolution of the grain.

EXAMPLES OF BSE APPLICATIONS

Albitization and K-Feldspar Replacement of Detrital Feldspars in Japan Sea Sandstones

Sandstones of Miocene age have been recovered from three sites in the Japan Sea backarc basin owing to deep drilling by the Ocean Drilling Program (ODP) (Boggs and Seyedolali, 1993). Sandstones at Site 796 in the northeast Japan Basin and Site 797 in Yamato Basin are volcaniclastic

Figure 4.19. Pore space (black) developed in a quartz grain owing to incipient grain dissolution. Authigenic anhydrite (1) and illite (2) partially fill some pore space. Rotliegend Sandstone, southern North Sea, depth 3500 m. Scale bar = 50 μm.

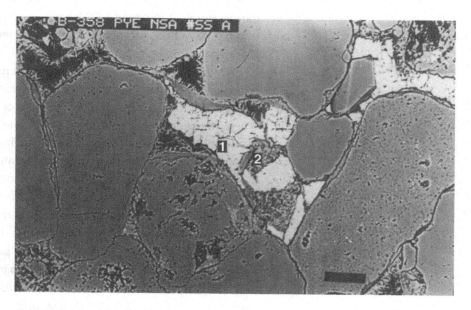

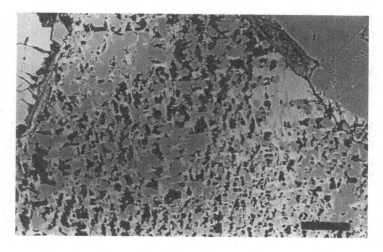

Figure 4.20 (left). Secondary porosity (black) formed in K-feldspar by selective dissolution of perthite. Carboniferous, southern North Sea, depth 3.7 km. Scale bar = 50 μm.

sandstones that contain abundant plagioclase feldspar and minor to moderate amounts of K-feldspar, which is mainly sanidine. Sandstones from Site 799 on Yamato Rise are feldspathic sandstones that contain abundant K-feldspar (orthoclase and microcline) and moderate plagioclase. Detrital feldspars in the sandstones at Sites 797 and 799, which were buried to depths of 800–1000 m and subjected to diagenetic temperatures exceeding 100°C, have been extensively albitized (Boggs and Seyedolali, 1992) and partially replaced by authigenic K-feldspar. Feldspars at Site 796, buried to a depth of only about 400 m and subjected to diagenetic temperatures of less than 65°C, have undergone no albitization but have been replaced to a minor extent by authigenic K-feldspar (see Fig. 4.12).

EDX and BSE image analyses show that many plagioclase feldspars from Sites 797 and 799 have been completely albitized. The most extensive albitization occurred at Site 797, where diagenetic temperatures within the sandstone sections may have risen as high as 150°C owing to intrusion by basaltic sills. Although diagenetic temperature clearly plays a major role in albitization, BSE image analysis shows that albitization within a given grain is also affected by the crystallographic characteristics of the grain and by the presence of features such as fractures, dissolution pores, and glass inclusions that allow fluid access to the grain. Many feldspar grains, both plagioclase and K-feldspar, from Sites 797 and 799 display evidence of selective albitization around such features (Figs. 4.21–4.23). BSE images reveal partially albitized areas with remarkable clarity because of the sharp contrast in gray-scale level compared to nonalbitized plagioclase or K-feldspar. For example, Figure 4.21 shows a partially albitized plagioclase grain. Note that albitized parts of the grain (large dark area and small dark patches) are clearly distinguished from the unaltered plagioclase. Figure 4.22 shows preferential albitization of a K-feldspar grain along cleavage or twin planes, and Figure 4.23 shows selective albitization around glass inclusions in plagioclase.

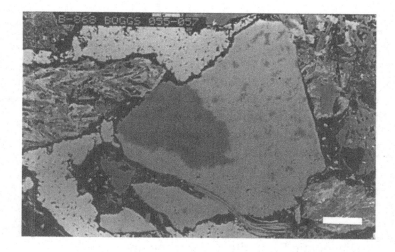

Figure 4.21. Large, partially albitized plagioclase grain (center of photograph). The dark patches are albite. Miocene sandstone, Ocean Drilling Program Leg 127, Site 797, depth 800 m below sea floor. Scale bar = 25 μm.

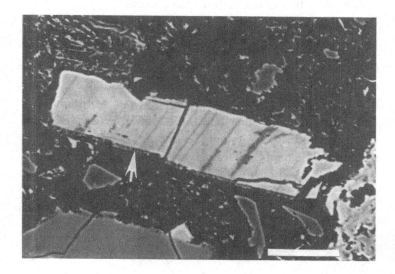

Figure 4.22. K-feldspar grain selectively albitized (dark lines) along cleavage or twin planes. Albite overgrowths (arrow) appear along the edge of the grain. Miocene sandstone, Ocean Drilling Program Leg 127, Site 797, Japan Sea. Scale bar = 20 μm.

Figure 4.23. Partial albitization of a plagioclase grain; albite (Ab) selectively replaces plagioclase around glass inclusions (black). Miocene sandstone, Ocean Drilling Program Leg 127, Site 797. Scale bar = 25 μm.

Many plagioclase feldspars from all three sites and some detrital K-feldspars from Sites 797 and 799 have been partially replaced by authigenic K-feldspar. This authigenic K-feldspar is optically clear, may be chemically pure (Or > 99), and occurs primarily as overgrowths on albite and detrital K-feldspar (Fig. 4.24). It may also occur in vein-like patches, replacing either albite or K-feldspar (see Fig. 4.12).

Diagenetic Evaporite and Carbonate Minerals in the Rotliegend Sandstone, Southern North Sea

The Rotliegend Sandstone (Permian) is an important gas reservoir rock in the southern North Sea Basin. The reservoir quality of the sandstone has been reduced through cementation by carbonate minerals, clay minerals, anhydrite, halite, and barite. On the other hand, dissolution processes have enhanced porosity by creating secondary pores owing to destruction of cements and framework grains.

By using BSE images, Pye and Krinsley (1986a) were able

Figure 4.24. Authigenic K-feldspar cement overgrowths (bright) on a partially albitized plagioclase grain. Miocene sandstone, Ocean Drilling Program Leg 127, Site 797. Scale bar = 25 μm.

to identify diagenetic minerals and secondary porosity and to work out the sequence of diagenetic events that affected reservoir quality of the Rotliegend Sandstone. For example, Figure 4.25 shows an oversized pore filled by late-state dolomite, siderite, and anhydrite cement. In Figure 4.26 an illitized detrital grain has been almost completely replaced by dolomite-ankerite-siderite rhombs. Figure 4.27 shows a secondary mouldic pore that formed by dissolution of quartz and that postdates surrounding carbonate and evaporite cement. Some detrital quartz grains have been partially dissolved, with subsequent infilling of some of the secondary porosity by authigenic illite and anhydrite (see Fig. 4.19).

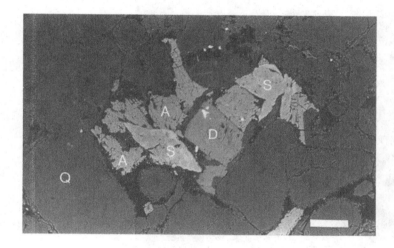

Figure 4.25. Oversized pore filled with late-stage, diagenetic dolomite (D), siderite (S), and anhydrite (A). Quartz grains (Q) show evidence of pressure solution at points of contact and are coated with authigenic illite. Rotliegend sandstone, southern North Sea. (*Source:* Pye and Krinsley, 1986a, fig. 3, p. 447.) Scale bar = 100 μm.

Figure 4.26. A detrital grain replaced almost completely by coalesced dolomite-ankerite-siderite rhombs (center of photograph). The large, dark grains are quartz. Rotliegend sandstone, southern North Sea. (From Pye and Krinsley, 1986a, fig. 11, p. 449.) Scale bar = 100 μm.

Figure 4.27. Secondary mouldic pore (3), formed by dissolution of quartz, that postdates surrounding carbonate and evaporite cement. 1 = siderite, 2 = ankerite, 4 = dolomite, 5 = anhydrite, 6 = late stage authigenic illite partly filling pore between quartz grains (dark). Rotliegend sandstone, southern North Sea. (From Pye and Krinsley, 1986, fig. 14, p. 453.) Scale bar = 100 μm.

Primarily on the basis of BSE image analysis, Pye and Krinsley interpreted three diagenetic stages during burial of the Rotliegend Sandstone: (1) alkaline, oxidizing conditions during shallow to intermediate burial; (2) acid, reducing conditions during intermediate to deep burial; and (3) alkaline, reducing conditions during deep burial and uplift. The transition from stage 1 to stage 2 was probably caused by expulsion of waters from the underlying Carboniferous shale. The transition to stage 3 probably began when faulting associated with uplift allowed invasion by alkaline fluids derived from Zechstein sediments.

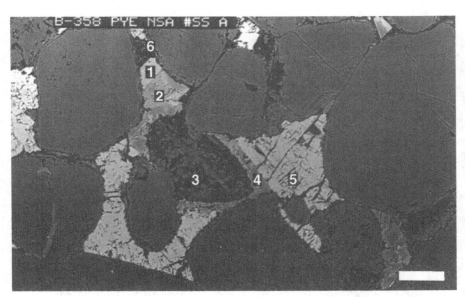

5

Shales

INTRODUCTION

Shales represent about one-half of all sedimentary rocks in the stratigraphic column. They consist mainly of clay minerals, quartz, feldspars, and micas, with minor amounts of other minerals such as carbonates, iron oxides, zeolites, and sulfates. Thus, they are mineralogically and chemically more complex than are either sandstones or limestones. Owing to their fine grain size, only a few petrographic studies of shales have been published. In thin section, many of the constituents of shales cannot be resolved optically because of their small size and the intermixing of clay minerals. Clay mineral flakes are so thin that several may be stacked irregularly upon one another so that light passing through is diffracted and/or refracted irregularly, producing an image with poor resolution. In addition, opaque materials such as hematite and amorphous inorganic and organic particles make petrographic viewing difficult (Blatt, 1982). Some workers have succeeded in using the petrographic microscope to study shales (e.g., Folk, 1960, 1962), but the process is time consuming and requires considerable expertise. The literature dealing with petrographic study of shales prior to 1980 is summarized by Potter et al. (1980). Recently, petrographic microscopy studies have been published by Schieber (1986, 1989, 1994), Odin (1988), Weaver (1989), and Leckie et al.

(1990), O'Brien and Slatt (1990), and Bennett et al. (1991a).

In spite of some success with petrographic methods, the mineralogy and texture of shales is difficult to determine petrographically. Shale mineralogy can be effectively studied by use of semiquantitative X-ray diffraction; however, the X-ray diffraction technique cannot be used to study texture, which is one of the most important characteristics of shales. The problem of poor resolution with the light microscope for textural studies has been solved in part by use of the scanning electron microscope (SEM) in the secondary electron mode (e.g., Barrows, 1980; Nuhfer, 1981; Nuhfer et al., 1981; O'Brien, 1984, 1985, 1987; O'Brien and Slatt, 1990); however, contacts between grains are very difficult to observe in the secondary mode.

The backscattered electron microscopy (BSE) method overcomes many of the problems encountered in optical petrography, X-ray diffraction, and secondary-mode SEM study of shales. It is an ideal tool for studying texture because textural relationships can be examined in great detail owing to high magnification. It is less useful for mineralogical analysis, but the mineralogy of many of the components can be determined in conjunction with X-ray analysis (EDX) and X-ray diffraction. A few workers have already applied the BSE technique to the study of shale textures (e.g., Krinsley et al., 1983; Pye and Krinsley, 1983, 1984, 1986b; Hugget, 1984; White et al., 1984; Pye et al., 1986; Burton et al., 1987; Prior and Behrmann, 1989; Brady and Krinsley, 1990; Krinsley et al., 1993). These published works are just the beginning of what promises to be an extremely productive method for studying the microfabrics of shales.

TEXTURES AND STRUCTURES

Microlaminations

The texture of shales encompasses grain size and shape, as well as microfabrics that result from the arrangement of clay particles and other grains (Bennett et al., 1991b). In addition, many shales contain sedimentary structures such as parallel lamination, convolute lamination, and trace fossils (e.g., mottled bedding). Laminae (layers < 1 cm thick) are of particular interest to geologists because of their potential environmental significance. Shales may also contain microfabrics generated by diagenetic chemical alteration or tectonic shear, which provide information about their postdepositional history.

Many shales are characterized by the presence of microlaminae that may be only a few grains thick and thus have a total thickness less than approximately 40 μm. Such microlaminae are difficult to observe with a conventional microscope but show up quite clearly in BSE. Shale microfabrics can range from those that display distinct single-grain layers (Fig. 5.1), to those with a more irregular kind of layering (Fig. 5.2), to those with no discernible layering (Fig. 5.3). Single-grain laminae (arrows, Fig. 5.1) or those that contain layers of only a few mineral grains may contain larger or smaller minerals of different composition intermixed with the platy minerals that define the laminae. In well-consolidated shales, microlayers and individual mineral grains bend around larger grains owing to differential compaction (Fig. 5.4).

Laminae are generated by suspension settling and also by some traction and turbidity currents. The presence or absence of laminae may be related to the oxygen state of the water (Moon and Hurst, 1984). Under anoxic conditions, clay particles tend to remain deflocculated and settle in a parallel arrangement conducive to formation of microlaminae, whereas clay particles that settle in well-oxygenated waters tend to clump together (flocculate) and settle in a random orientation. Thus, organic-rich black shales, which form under anoxic conditions, are characterized particularly by the presence of laminae and microlaminae (Fig. 5.5), whereas many gray shales deposited in oxygenated waters do not. Absence of laminae may also be the result of bioturbation by organisms.

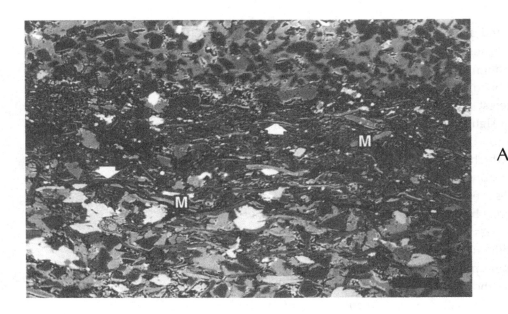

A

Figure 5.1. BSE photographs (A) and (B) show microlaminae (arrows) formed by parallel arrangement of single mica flakes (M). Other minerals include iron oxides (i), pyrite (P), carbonate grains (C), quartz (Q), and smaller clay minerals (probably illite and kaolinite) oriented more or less parallel to larger clay-mineral flakes. (B) shows some topographic relief, caused by adding the secondary to the backscattered signal. Core samples from the Lower Tertiary, Tasman Sea, Tasmania. Scale bars: (A) = 100 μm, (B) = 50 μm.

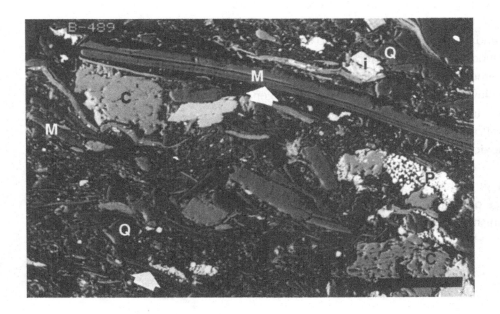

B

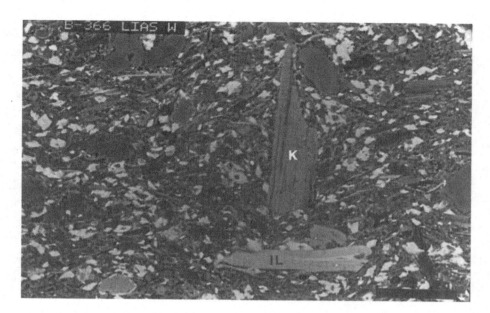

Figure 5.2 (above). BSE photograph showing less regular layering than that in Figure 5.1. The large flakes are kaolinite (K) and illite (IL). The lighter layer in the illite flake is richer in iron than the remainder of the grain. Finer-grain minerals in the background are mainly quartz, feldspar, and authigenic clay minerals. Whitby Mudstone, Lower Jurassic, Yorkshire, England. Scale bar = 50 μm.

Figure 5.3 (below). BSE photograph showing random orientation of clay minerals and other grains within a shale. The absence of microlamination is probably the result of bioturbation. A large clay mineral (CM) has slightly penetrated a quartz grain (Q), which displays a corroded edge owing to dissolution. Note the quartz (Q) grain that forms a nucleus for the large, rounded glauconite (G) grain. Whitby Mudstone, Lower Jurassic, Yorkshire, England. Scale bar = 10 μm.

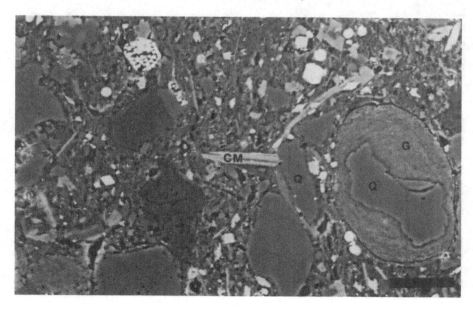

77

Figure 5.4 (above). Micaceous shale, showing deformation of micas owing to compaction. Other minerals include kaolinite (K), some of which is authigenic, muscovite (M), quartz (Q), and pyrite (P). Grey Shale Member, Whitby Mudstone Formation (Lias), Yorkshire, England. Scale bar = 10 μm.

Figure 5.5. Calcite-rich lenses (white) and organic matter (black) in an oil shale. The calcite-rich lenses, which may be flattened fecal pellets, contain fragments of coccoliths that are partly neomorphosed. The small white grains (arrows) are pyrite framboids. The well-preserved lamination indicates anoxic bottom conditions with minimal bioturbation, Kimmeridge clay, East Anglia, England. Scale bar = 100 μm.

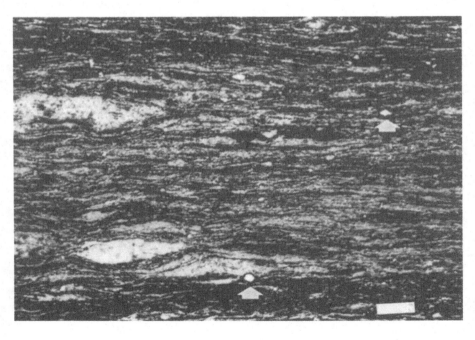

Microstructure of Individual Clay Minerals and Micas

The high magnification possible with BSE allows detailed study of the microstructure of individual mineral grains as small as 1–5 μm. Photographs of clay minerals and micas taken at magnifications of 2,000–20,000 times can reveal the presence of multiple mineral layers within individual grains and intergrowths of other minerals between these layers (Figs. 5.6 and 5.7). For example, kaolinite or iron oxides may crystallize and grow between mica layers, causing individual grains to expand (Fig. 5.7). Hydrous layer silicates measuring 1–5 μm are probably authigenic; those larger than 5–10 μm are likely detrital micas such as biotite and muscovite. Features of detrital grains such as the broken ends of the mica flake in Figure 5.7 can help differentiate detrital from authigenic grains.

Microfabrics in Metamorphosed Shales

BSE is also proving to be useful in the study of microfabrics in shales that have been altered by metamorphic pro-

Figure 5.6. Highly magnified BSE image of a mixed-layer clay mineral. The dark gray parallel layers are kaolinite; the light material is illite. Black layers are cracks. The diffuse, black stripes perpendicular to the layers within the illite represent initial diagenetic change resulting from removal of some iron and manganese. Lower Jurassic, Whitby Mudstone, Yorkshire, England. Scale bar = 5 μm.

Figure 5.7. A large sericite flake (center of photograph), broken at both ends, containing authigenic kaolinite (gray) between layers. The black layers are cracks. Small, bright iron oxide crystals are present between the sericite layers. Broken ends suggest that the original flake was detrital. The quartz (Q) grain in the right side of the photograph is partially dissolved. Lower Jurassic, Whitby Mudstone, Yorkshire, England. Scale bar = 10 μm.

cesses. For example, White et al. (1985) demonstrated that BSE is a powerful technique for studying intergrowths in phyllosilicates and can also reveal compositional zoning in a given intergrowth phase. Figure 5.8 shows a large detrital chlorite (with compositional zoning) in a matrix of authigenic chlorite and illite, and Figure 5.9 shows intergrowths

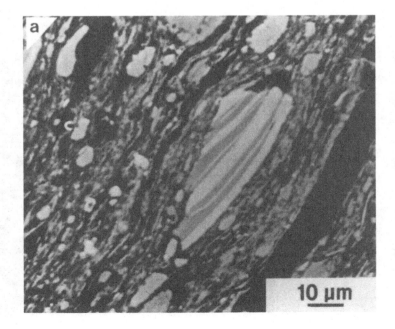

Figure 5.8. General microstructure of large detrital chlorite grain in a fine-grained matrix consisting mainly of incoherently intergrown chlorite and illite. Lothian Oil Shale (Carboniferous), Scotland. Scale bar = 10 μm. (From White et al., 1985; reproduced by permission.)

Figure 5.9. Microstructure of coarse, chlorite/phengite intergrowths in schists from the chlorite subzone. Haast Schists, New Zealand. The dark grains are quartz and albite. Epidote grains are marked E. Scale bar = 50 μm. (From White et al., 1985; reproduced by permission.)

of chlorite and phengite (muscovite). The studies of White et al. demonstrate that intergrowths of illite, chlorite, and kaolinite are common in both detrital and authigenic pore-filling phases and range in scale from very fine (i.e., 0.25 μm thick) to much broader lamellae several microns wide. Thus, both coarse (detrital) and fine (authigenic) clay minerals can be studied in detail with BSE, providing information about both detrital and authigenic phases.

BSE is also highly effective for studying tectonic fabrics that result from deformation processes. An example from a Jurassic (?) black shale of the Argille Scagliose of the Italian Apennines (Agar et al., 1989) is shown in Figure 5.10. The microfabric of this tectonically altered shale is strikingly defined in the BSE photograph.

Particle Size and Shape

BSE provides a means for studying both the size and shape of shale grains that are too small to be viewed effec-

Figure 5.10. Montage of BSE photographs from core of macroscopic fold of Argille Scagliose, Jurassic to Oligocene, Italian Apennines. S: earlier bedding-parallel fabric is overprinted by strong axial planar fabric (A). P: pods in which earlier microfabric is preserved. Microfabric is defined by alignment of quartz (Q) and calcite (C) long axes, particularly tests of forams (F). Small grains impinge into some calcite clasts. Phyllosilicate fabric is generally aligned with long axes. Phyllosilicates are buckled around clasts (B) and folded and kinked into spaces between clasts (K). Small, round bright grains are pyrite framboids. Black areas are epoxy-filled pore space. Scale bar = 50 μm. (From Agar et al., 1989; reproduced by permission.)

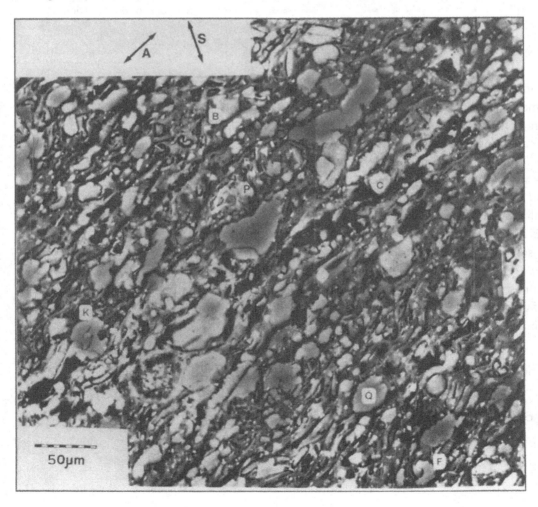

tively with a petrographic microscope. The two-dimensional sizes of particles that are less than 1 μm can be measured in BSE images, allowing both the estimation of the mean size and the sorting of shale particles. Textural variability is generally high in shales, and bimodality of grain size is a common feature. Many shales contain a coarse (10–60 μm) fraction, which is probably detrital, and a finer fraction (10–1 μm or less) that is likely authigenic. Figure 5.3 shows a bimodal shale with large quartz and clay mineral grains together with much finer-grained minerals that may include clay minerals, quartz, feldspars, and pyrite. If examined at very high magnification, the fraction of shales measuring less than 1 μm commonly has a somewhat feathery appearance (Fig. 5.11). This very fine material is likely of authigenic origin, generated during diagenesis by chemical alteration of feldspars or other minerals. Figure 5.12 also demonstrates the bimodal size distribution of grains in a shale.

The shapes of grains, including grains smaller than 1 μm, can also be studied very effectively in BSE images. The shape of small grains is probably not a useful indicator of grain transport mode because the shapes of fine particles – particularly quartz grains, which are the grains commonly ana-

Figure 5.11. Authigenic clay minerals, ranging in size from ~25 μm to less than 1 μm displaying a feathery appearance. Core sample from the lower Tertiary, Tasman Sea, Tasmania. Scale bar = 10 μm.

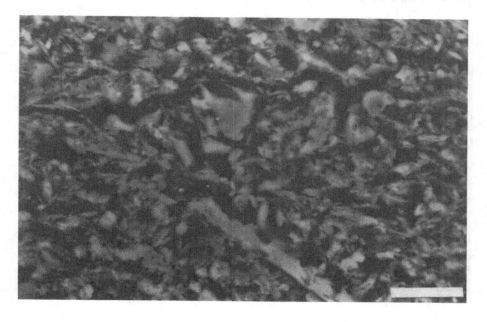

lyzed in shape studies – are not changed significantly during transport. Kuenen (1960) demonstrated, for example, that quartz grains smaller than about 0.05 mm do not become rounded during transport, even by eolian action. On the other hand, the shapes of particles can be a useful criterion for differentiating between detrital and authigenic grains in shales. The presence of broken ends on soft clay minerals or mica flakes, such as that in Figure 5.7, is indicative of detrital origin. By contrast, authigenic minerals either have well-developed, euhedral, crystal faces or display irregular, intricate intergrowths among other crystals. Authigenic quartz, feldspar, and kaolinite, for example, commonly display well-developed crystal faces, whereas some other minerals (such as the pyrite in Figure 5.1B) have shapes so irregular that they must have grown in place. See also the discussion below under authigenic minerals.

FOSSILS

Fossils are important constituents of many shales. BSE is most useful for studying the details of microfossils and fecal

Figure 5.12. Coarse and fine phases in a silty shale. Large quartz grains (Q) are rimmed with illite or chlorite. Both fine clays and coarser illite/chlorite clay minerals in the fine phase are oriented. C = carbonate (dark Mg-rich; light Fe-rich); P = pyrite. Core sample from lower Tertiary sediments, Tasman Sea, Tasmania; 37 m below sea floor. Scale bar = 100 μm.

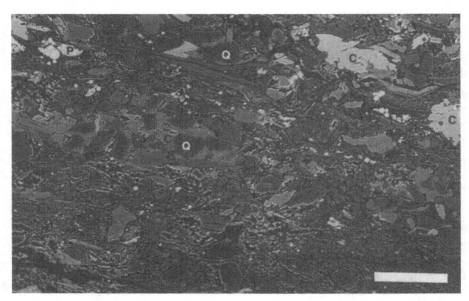

pellets that are too small for effective viewing with a petrographic microscope. Both the shapes and mineralogy of microfossils can be altered by diagenetic processes. Analysis of these diagenetic changes reveals the timing and sequence of cementation and replacement events that occur during the burial of shales. Several examples from the Texas Gulf Coast illustrate how BSE images can be used to study and interpret these diagenetic events.

Figures 5.13–5.16 show foraminifers in Miocene sediments recovered from a Mobil Oil Company well drilled near Galveston, Texas. Because they are composed of calcite, foraminifer tests are more resistant to diagenetic alteration than are the shells of organisms composed of aragonite or Mg-calcite. Nonetheless, foraminifer tests are altered in various ways during burial. In Figure 5.13, the test is still composed of the original calcite, but the chambers are almost filled with authigenic calcite (cement). Clay minerals partially replace calcite cement in the large chamber. The foraminifer test and several chamber walls in Figure 5.14 are almost completely replaced by kaolinite, indicating that this specimen has been subjected to an episode of carbonate dissolution, with concomitant clay mineral replacement. The

Figure 5.13. Planktonic foraminifer in shale. The chamber walls are composed of original calcite and display the presence of large pores; interiors of the chambers are almost completely filled with calcite cement. Dark areas within the chamber filling are solution pores. The surrounding shale is composed of kaolinite, quartz, feldspars, and calcite. Several pyrite framboids and isolated pyrite crystals (very bright areas) are present in the right side of the photograph. Miocene, Mobil Oil Co. well near Galveston, Texas. Depth 2000–2100 m. Scale bar = 20 μm.

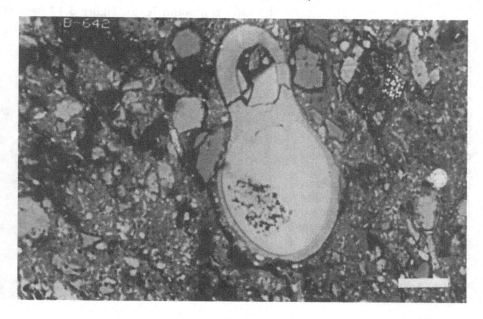

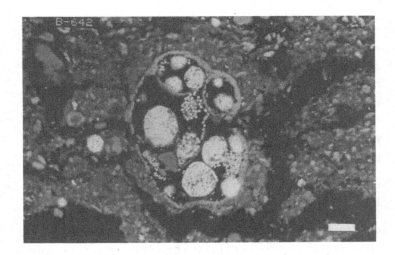

Figure 5.14. Planktonic foraminifer in which the test wall and several chamber walls have been replaced by kaolinite. Authigenic pyrite framboids are present inside the test chambers. The surrounding shale consists of kaolinite, quartz, feldspars, pyrite, calcite and Mg-calcite. Miocene, Mobil Oil Co. well near Galveston, Texas. Scale bar = 10 μm.

chambers of the test are partially filled with framboidal pyrite. The foraminifer in Figure 5.15 has undergone extensive recrystallization, which has destroyed original shell structure, producing a massive-appearing test wall. The chambers are partially filled with clay minerals and calcite. Figure 5.16 shows a foraminifer that has been completely replaced by chlorite/illite at a burial depth of 3400 m.

Figures 5.17 and 5.18 show foraminifers in Miocene/ Oligocene sediment recovered from the Kerlin Well (Magnolia Oil Co.), South Texas. Figure 5.17 depicts a foraminifer

Figure 5.15. Planktonic foraminifers in shale. The test in the middle right side has been recrystallized to massive calcite, but the outline of the test is still visible. Note the foraminifer ghost at the bottom center. Next to the larger test is a single crystal of calcite with a darker core of magnesian calcite. The material surrounding the tests is mostly kaolinite and calcite, which has a composition and size distribution slightly different from the filling within the large foraminifer. Miocene, Mobil Oil Co. well near Galveston, Texas. Scale bar = 10 μm.

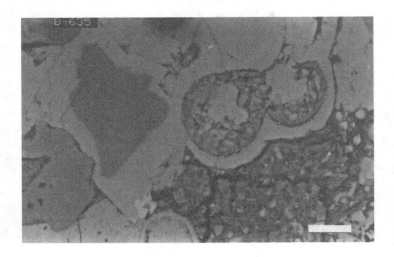

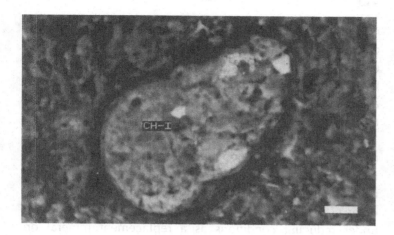

Figure 5.16. Foraminifer completely replaced by chlorite/illite (CH-I). The test walls, where visible, are replaced by kaolinite. The interior of the test contains chlorite, illite, calcite, and incipient pyrite framboids. Material surrounding the test contains more kaolinite and quartz than does the test interior. Note that this kind of information could not be obtained by conventional paleontologic techniques because separating the test from the shale would cause it to disintegrate. Miocene, Mobil Oil Co. well near Galveston, Texas. Depth 3660–3666 m. Scale bar = 10 μm.

from a depth of 300 m. The shell material has been partially replaced by kaolinite, and the chambers are partially filled with iron-rich calcite. The foraminifer in Figure 5.18, 2900 m deep, has undergone recrystallization; otherwise, the test is intact. The chambers are filled with calcite cement, partially replaced by pyrite.

These diagenetic alterations reflect changes in the chemical burial environment, although it is difficult to relate the changes directly to increasing burial depth. Increases in temperature with increasing burial depth favor recrystallization of carbonate minerals and the precipitation of carbonate ce-

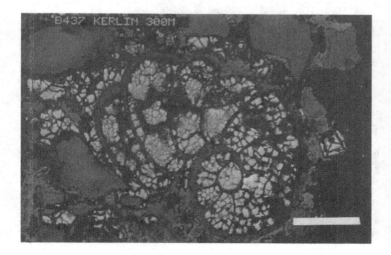

Figure 5.17. Planktonic foraminifer in a silty shale. The shell material has been replaced by kaolinite. The inside is partially filled with Fe-rich calcite, which appears very light owing to the SEM contrast and brightness settings. The large dark grains are quartz. Miocene/Oligocene, Kerlin Well (Magnolia Oil Co.), South Texas. Scale bar = 10 μm.

ments; however, dissolution of carbonate tests and cements can also occur under higher-temperature conditions (see, e.g., Fig. 5.13, depth of ~2100 m). Carbonate dissolution probably results from an increase in pore water pH owing to an episode of decarboxylation, associated with shale dewatering, that generates carboxylic acids (e.g., Surdam et al., 1989). The formation of clay mineral cements such as chlorite and illite is a common diagenetic process, in which the ions necessary for clay-mineral formation are probably furnished by chemical decomposition of feldspars, micas, and other clay minerals (e.g., smectite, kaolinite). Pyrite forms under reducing conditions as a replacement mineral or cement.

AUTHIGENIC MINERALS

The mineral composition of shales varies widely; however, the average shale contains about 50% clay minerals, 30% quartz, 10% feldspars, and 10% other minerals, such as carbonates, pyrite, iron oxides, and zeolites (Boggs, 1992, p. 281). Many of the minerals in shales are authigenic because

Figure 5.18. Planktonic foraminifer in a carbonate-rich phase of a silty shale. The foraminifer in the upper left is composed of calcite that has been almost completely recrystallized. Dark gray grains are quartz. Note pyrite (bright) within chambers. Miocene/Oligocene, Kerlin Well (Magnolia Oil Co.), South Texas. Scale bar = 10 μm.

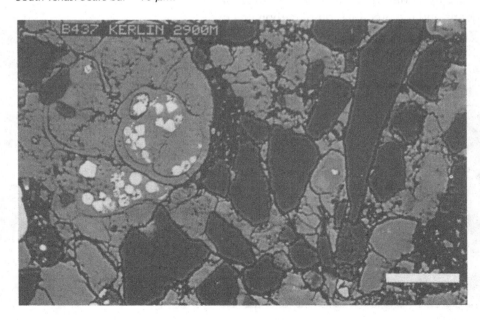

burial of shales in subsurface environments (characterized by higher temperatures and pressures, changed pH and Eh conditions, and pore waters of different composition from those of the depositional environment) can bring about significant changes in their mineral composition. Smectite clay minerals are commonly altered to illite or chlorite, and kaolinite may be destroyed. Feldspars and carbonate minerals may also be destroyed by dissolution under some conditions. Under other conditions, authigenic feldspars, quartz, carbonate minerals, glauconite, and pyrite are formed. Authigenic minerals such as carbonates, quartz, and pyrite may be present as cements or they may replace other minerals. Many of these authigenic mineral phases can be easily identified in BSE images. Some additional examples of authigenic minerals will serve to amplify this discussion.

Carbonates and Sulfates

Most carbonate minerals in shales are present as fossils (Fig. 5.13) or as authigenic growths of some kind. For example, Pye (1985a) describes authigenic zoned dolomite rhombs in black shales of the Jet Rock Formation of northeast England. These rhombs, 5–20 μm in diameter, have various chemical compositions but consist mainly of a dolomite core surrounded by a zone of ferroan dolomite, ankerite, or ferroan calcite. Figure 5.19 shows complex carbonate zoning involving normal (low-magnesian) calcite, high-magnesian calcite, and ferroan calcite. All the carbonate rhombs display similar patterns of zoning but differ in detail. Such variations are likely the result of differences in microenvironmental conditions on a micron scale. Textural evidence presented by Pye suggests that dolomite formation occurred at shallow burial depths by direct precipitation from pore fluids within the sulfate reduction zone. Figure 5.20 shows authigenic siderite crystals, formed by replacement of silicate minerals in a micaceous shale. Variations in Mg content of carbonate

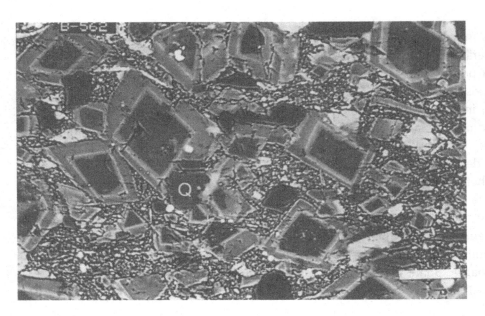

Figure 5.19. Chemically zoned carbonate rhombs. The dark centers are rich in Mg; the very bright rims around the centers are Fe-rich. The thick, gray outside rims have the chemical composition of normal calcite. The rounded, black grains are quartz (Q); the small, round, bright grains are pyrite. The larger grains are embedded in a fine-grained matrix consisting of clay minerals, feldspar, quartz and minor iron oxide. Carbonate phase of black shale, Ordovician, Beechcraft Mountain, New York. Scale bar = 50 μm.

Figure 5.20. Authigenic siderite crystals (gray) that grew replacively in micaceous mudstone. The small, bright, rounded grains are pyrite framboids. Alum Shale Member, Whitby Mudstone Formation (Lias), Yorkshire, England. Scale bar = 100 μm.

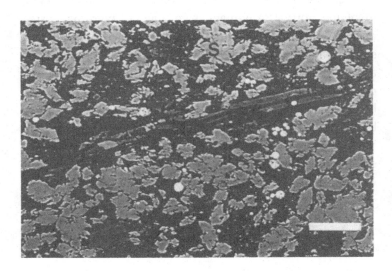

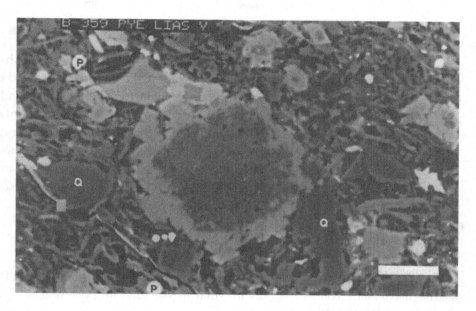

Figure 5.21. Large zoned calcite crystal (center of photograph) with a low-Mg calcite rim and a high-Mg calcite core. Note the presence also of quartz (Q) and pyrite (P). Whitby Mudstone Formation (Lias), Yorkshire, England. Scale bar = 20 μm.

crystals in shales are revealed with great clarity in BSE photographs. The rim of the large carbonate crystal in Figure 5.21 consists of low-Mg calcite (< 4 mol % $MgCO_3$), whereas the core consists of high-Mg calcite. Note the presence also of quartz and pyrite in this shale. Authigenic gypsum or anhydrite may be present in shales where oxidation of pyrite furnishes a source of sulfur (Fig. 5.22).

Figure 5.22. Weathered pyritic shale, showing growth of gypsum crystals (G) between partially opened shale laminae. Oxidation of pyrite by the bacterium Thiobaccillus ferrooxidans produces sulphuric acid, which reacts with calcite to form secondary gypsum. Growth of the gypsum causes a volume expansion that ruptures the lamination, thereby accelerating the pyrite oxidation process. P = pyrite. Edale Shale (Carboniferous), Derbyshire, England. Scale bar = 10 μm.

Micas and Clay Minerals

Although some of the clay minerals in shales are detrital, smectite and kaolinite tend to be destroyed during diagenesis with concomitant formation of illite and chlorite. Thus, older, more deeply buried shales commonly contain abundant authigenic illite and chlorite. For example, Figure 5.23 shows a large chlorite stack with illite lamellae, and Figure 5.24 shows bands of authigenic chlorite-mica. Authigenic kaolinite (Fig. 5.25) may form in shales also, particularly at temperatures below approximately 25°C.

Pyrite

Almost all pyrite in shales is authigenic. Pyrite may occur as irregular masses that fill pore spaces as a cement or replace other minerals or organic particles (Figs. 5.4 and 5.21) or it may occur as framboids (microscopic aggregates of pyrite crystals in spheroidal clusters resembling raspberry seeds). Figure 5.26 shows clusters of pyrite framboids in a shale, and Figures 5.27 and 5.28 demonstrate the intricate structure of a pyrite framboid as revealed in a high-magnification BSE photograph. The formation of pyrite requires reducing conditions, and it is most common in organic-rich shales.

Figure 5.23. Large chlorite grains, consisting of stacks of chlorite and illite (darker layer), in a shale that has undergone partial recrystallization during very low-grade metamorphism. Chlorite-mica stacks may have more than one mode of origin, but many appear to have originated as diagenetically altered detrital biotite grains. Note the apparent 3-D topography, created by mixing the BSE and SE signals. The topography has been enhanced and the BSE signal somewhat degraded. Upper Ordovician black shale near Aberangell, Gwynedd, Wales, U.K. Scale bar = 50 μm.

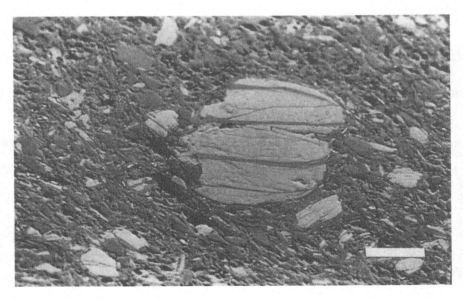

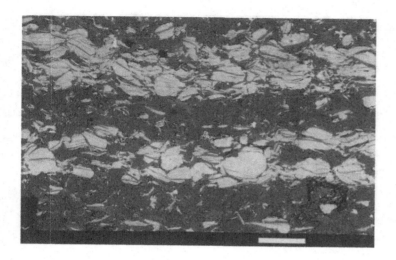

Figure 5.24. Parallel bands (bright) consisting of chlorite-mica stacks in a muddy siltstone. The 001 planes of the chlorite are parallel to laminations in the siltstone. Upper Ordovician, Central Wales. Scale bar = 100 μm.

Figure 5.25. Authigenic kaolinite macrocrystals (arrow, K) containing small pyrite framboids (P). Q = quartz; M = mica. Jurassic (Lias) shale, southern North Sea, depth 3500 m. Scale bar = 10 μm.

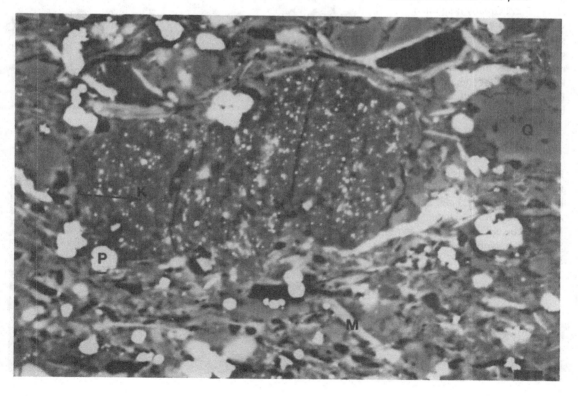

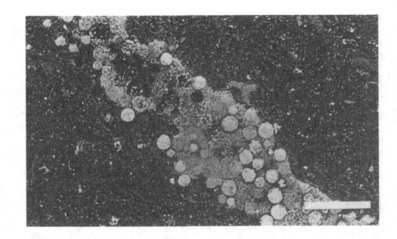

Figure 5.26. Pyrite framboids (bright spheres) scattered through a shale from the Adventdalen Group (Upper Jurassic/Lower Cretaceous), central Spitsbergen, Norway. Many of the smaller pyrite crystals present among the framboids are not stoichiometric pyrite and may represent framboids in the process of formation. Scale bars = 50 μm.

Figure 5.27. Greatly magnified view of a pyrite framboid in a shale. Note the surrounding fine-grained matrix, which consists mainly of clay minerals (mostly kaolinite). Oligocene, Frio Formation, Texas Gulf Coast. Scale bar = 20 μm.

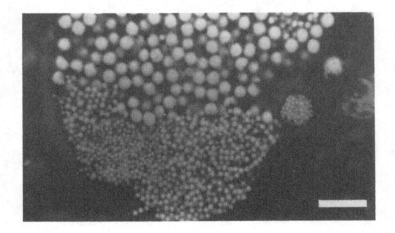

Figure 5.28. Three sets of pyrite framboids of contrasting sizes in a highly organic-rich shale. Adventdalen Group (Upper Jurassic/Lower Cretaceous), Central Spitsbergen, Norway. Scale bar = 5 μm.

Glauconite

Glauconite is a common authigenic mineral in sandstones and occurs also in some shales. For example, Dypvik et al. (1992) report glauconite from Cretaceous shales in Spitsbergen, Norway. The glauconite grains float in a feathery matrix consisting of illite/smectite clays (Fig. 5.29). Within certain horizons, the glauconite grains are mainly angular in shape with sharp boundaries in contact with surrounding pyrite and the matrix. Other examples of glauconitic shales include the Lower Tertiary shales of the Tasman Sea and the Lower Jurassic Whitby Mudstone, Yorkshire, England (Krinsley et al., 1993). Most glauconite grains in these shales are well rounded and contain many different kinds of internal structures, including well-crystallized quartz, feldspars, and clay minerals, which apparently acted as nucleation centers (see Chapter 8).

PRIMARY AND SECONDARY POROSITY

BSE provides an effective means of studying porosity in siliciclastic and carbonate rocks. It is especially useful for

Figure 5.29. Large, subangular, poorly sorted glauconite grains, with a fine-grained interstitial matrix consisting of illite/smectite clays. Myklegardfjellet (clay) Bed (Cretaceous), Spitsbergen, Norway. Scale bar = 100 μm.

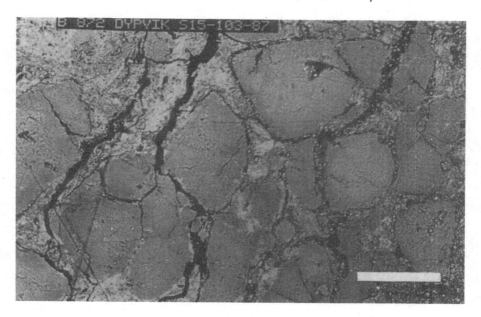

Figure 5.30. Porosity in a shale, identified in this BSE image by the black areas among grains. Both intercrystalline and intracrystalline porosity are present; however, most of the porosity is intracrystalline. The pores have random shapes except those generated by dissolution of minerals with specific shapes. (B) is a closeup of (A). Upper Ordovician shale, Wales, U.K. Scale bars: (A) = 10 μm; (B) = 10 μm.

A

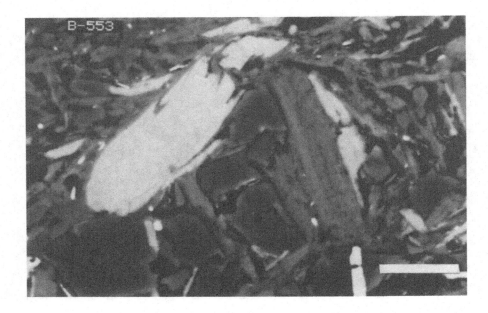

B

studying the microporosity of shales, in which pores are much too small to be seen with an ordinary light microscope. Specimens for study must be prepared carefully (e.g., by impregnating with epoxy before grinding) to ensure that the pores are not the result of plucking during grinding. Also, the presence of fine-grained organic matter can cause problems because both organic matter and pores appear dark, and organic matter can fill holes. With care, however, pore space in shales can be easily identified in BSE images and is subject to computer analysis (Tovey and Krinsley, 1991). The shape, average size, and orientation of pores can all be determined. Figure 5.30 provides an example of the microporosity in a shale at high magnification.

6

Carbonates

INTRODUCTION

Carbonate rocks are generally less complex mineralogically than are siliciclastic sedimentary rocks. Ancient limestones are composed mainly of calcite, although some dolomite and aragonite (in very young rocks) may also be present. Low-magnesian calcite is the typical form of calcite; high-magnesian calcite (> 4 mol % $MgCO_3$) is less common. Ancient dolomites (dolostones) are composed mainly of the mineral dolomite, with perhaps small amounts of low-magnesian calcite. Other carbonate minerals such as magnesite, siderite, ankerite, and strontionite may be present in carbonate rocks in very minor amounts. The mineralogy of carbonate rocks is commonly studied by petrographic (light) microscopy, after etching and staining carbonate thin sections with an appropriate stain (e.g., Miller, 1988a). Carbonate minerals can be differentiated also by X-ray diffraction methods or, less commonly, by electron probe microanalysis (EPMA).

Texturally, limestones consist of various kinds of carbonate grains (clasts, fossils, ooids, peloids), microcrystalline calcite (lime mud or micrite), and sparry calcite cement. Identification of these textural elements is often of greater interest to geologists than is simple mineral identification. In addition to petrographic and X-ray diffraction analysis, the study of carbonate minerals has been greatly enhanced by use of cathodoluminescence (CL) techniques (e.g., Miller, 1989b; Barker and Kopp, 1991). CL petrography has proved to be especially useful in distinguishing among low-magnesian calcite, high-magnesian calcite, and dolomite, and in the study of zoning in dolomite crystals and carbonate cements.

In contrast to optical microscopy and CL petrography, application of BSE imagery to the study of carbonate rocks is still in the incipient stages. Only a few carbonate workers are now beginning to use BSE, commonly in conjunction with CL studies, and to publish BSE photographs of carbonate rocks. Nonetheless, the potential applications of BSE to the study of carbonate rocks are enormous. The greater magnification and resolution of BSE compared to CL and standard light microscopy makes BSE particularly useful for studying fine-grained carbonates and the details of crystals such as zoned dolomite crystals and zoned carbonate cements. BSE also has important applications in the study of porosity and microporosity and for examining detailed grain-to-grain or crystal-to-crystal relationships.

APPLICATIONS

Possible applications of BSE imagery to the study of carbonate rocks include analysis of original depositional textures (e.g., characteristics of carbonate grains), structures (e.g., cryptalgal fabrics, stromatactis), and secondary features generated during diagenesis. The BSE method is particularly well suited to the study of diagenetic changes brought about by both physical and chemical diagenetic processes: pressure solution, cementation, dolomitization and other replacement phenomena, and dissolution.

Zoned Dolomite

A particularly appropriate application of BSE to the analysis of carbonate rocks is the study of zoning in dolomite crystals. Many rhombic dolomite crystals exhibit a cloudy, rhombic central zone surrounded by a clear or nearly clear rim. The clear rims are secondary, syntaxial rims that formed by precipitation into pore space around a cloudy center, possibly pore space created by dissolution of $CaCO_3$ from just beyond the limits of the cloudy rhombs. In addition to these clear syntaxial rims, many dolomite crystals display fine-scale internal zoning that results from differences in composition, particularly iron content. Ferrous iron is common in many dolomite crystals as a substitute for magnesium. Reeder and Prosky (1986) attribute this type of zoning to systematic compositional differences between nonequivalent growth sectors in dolomite rhombs and refer to it as **sector zoning**. They report that growth sectors forming under {1120} crystal faces are enriched in Fe and Mg and are slightly enriched in Mn relative to sectors forming under {1014} faces.

Unless ferrous iron is subsequently oxidized to ferric iron (hematite), the sector zoning in dolomite crystals is not visible under a petrographic microscope, although staining can sometimes reveal the zoning. Therefore, zoning in dolomite has commonly been studied in the past by CL petrography. BSE can be an important supplement to CL petrography in studying compositional zoning (Amthor and Friedman, 1992; Amthor, 1993) and may show details of zoning not visible, or poorly visible, by CL. In particular, zones within a dolomite crystal that contain a very high iron content are nonluminescent under CL. These zones show up clearly in BSE. For example, the CL image in Figure 6.1 shows fine-crystalline host dolomite overgrown by zoned dolomite cement, followed by a later-stage void-filling dolomite. This late-stage dolomite is nonluminescent owing to its high iron content. A BSE image (Fig. 6.2) of the zoned crystal in Figure 6.1 sharply reveals a bright rim on the zoned dolomite that is not visible in the CL image. This bright rim has a higher iron content than the void-filling dolomite, hence the contrast in gray scale between the rim and the void-filling cement. Fig-

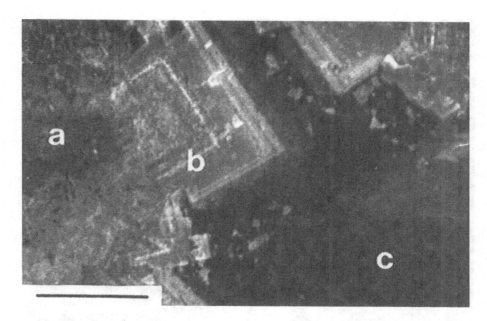

Figure 6.1 (above). Cathodoluminescence photograph of a zoned
dolomite crystal. Dolomite (a) was
discontinuously overgrown by
zoned dolomite cement (b), which
in turn was followed by nonplanar
dolomite (c), occluding porosity.
Ellenburger Group (Ordovician),
Permian Basin, West Texas. Scale
bar = 0.2 mm. (From Amthor and
Friedman, 1992, fig. 5D, p. 136;
reproduced by permission.)

Figure 6.2 (below, left). BSE image
of the zoned crystal in Figure 6.1.
Fe content of this crystal increases
outward in the rim, resulting in a
very bright BSE zone and a dull to
nonluminescent cathodoluminescence interior. The nonplanar
dolomite (D) surrounding the
zoned crystal has lower Fe concentrations and a slightly darker BSE
color. Note the scalloped, dissolution contact (arrows) between the
zoned and surrounding dolomite.
Ellenburger Group (Ordovician),
Permian Basin, West Texas, Scale
bar = 100 μm. (From Amthor and
Friedman, 1992, fig. 6A, p. 137; reproduced by permission.)

Figure 6.3 (above). BSE image of zoned dolomite crystals that display an abrupt, sharp boundary between iron-poor dolomite in the center of the crystals (dark) and iron-rich dolomite rims (bright). Ellenburger Group (Ordovician), Permian Basin, West Texas. Scale bar = 0.1 mm. (From Amthor and Friedman, 1992, fig. 5E, p. 136; reproduced by permission.)

Figure 6.4. BSE image showing mottled dolomite matrix that grades toward the top of the photograph into pore-filling (rim) dolomite containing patches of near-stoichiometric dolomite (dark) in an overall Ca-rich matrix (bright). Ellenburger Group (Ordovician), West Texas. Scale bar = 100 μm. (From Kupecz and Land, 1991, fig. 17, p. 564; reproduced by permission.)

ure 6.3 shows an additional example of a zoned dolomite crystal that displays a razor-sharp boundary between the iron-poor (dark) interior of the crystal and the iron-rich rim (bright). Figure 6.4 is a BSE image of a dolomite crystal that shows mottled matrix dolomite, which grades into pore-filling dolomite (bright), containing patches of near-stoichiometric dolomite (dark BSE image) in an overall calcium-rich matrix (bright BSE image). The relationship between BSE image and stoichiometry was established by electron microprobe analysis.

An additional example of a zoned crystal is shown in Figure 6.5. This crystal, near the center of the photograph, has a dark core consisting of near-stoichiometric dolomite. The surrounding, brighter outer zone consists in part of high-Mg calcite and in part of iron-rich dolomite (ankerite). The zoned crystal partially replaces a foraminifer test and other calcite crystals.

Figure 6.5. Zoned carbonate rhomb (arrow) partially replacing a recrystallized foraminifer test (F) and other calcite crystals (gray). The dark core of the rhomb consists of dolomite; the outer, brighter zone consists in part of high-Mg calcite and in part of Fe-rich dolomite (ankerite). The dark grains are quartz (Q). Mobil Oil Company Halls Bayou Ranch #1 Wildcat Well, Galveston County, Texas. Scale bar = 10 μm.

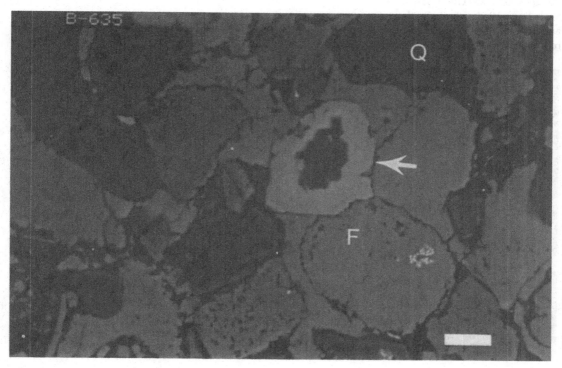

Carbonate Cements and Cement Stratigraphy

Owing to the outstanding resolution of the BSE method, fine- and coarse-grained intergranular carbonate cements or cements within fossils or other carbonate grains can be studied very effectively by BSE. BSE not only reveals textural details that may not be visible by optical or CL petrography but also, in conjunction with EPMA, allows the mineralogy of the cement to be established. Several examples illustrate the method. Figure 6.6 shows complexly zoned crystals of calcite, dolomite, and ankerite growing inward and outward from the test of a foraminifer. Figure 6.7 shows complexly zoned rhombs within a foraminifer test that consist of a core of dolomite, followed by zones of calcite and ankerite. Figure 6.8 shows a foraminifer with macroscopic biomoldic porosity (black), low-Mg calcite cement (white) in intraparticle chambers of the foraminifer, and the undissolved shell wall (gray). Finally, in Figure 6.9, calcite cement infills algal interconnected tubules in a shallow-marine limestone.

During the progressive burial and diagenesis of a limestone, more than one generation of cement may be precipi-

Figure 6.6. Carbonate crystals and complexly zoned rhombs of calcite, dolomite, and ankerite both inside and outside a foraminifer test. Q = quartz; F = foraminifer test. Frio Formation (Oligocene), Texas Gulf Coast. Scale bar = 10 μm. (From Brady and Krinsley, 1990, fig. 5, p. 5; reproduced by permission.)

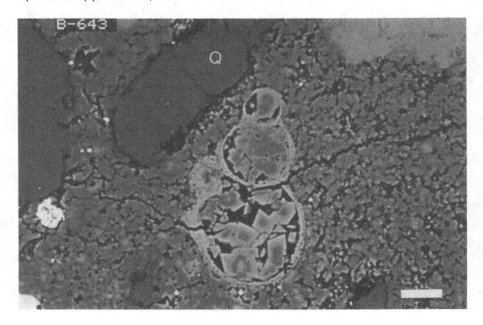

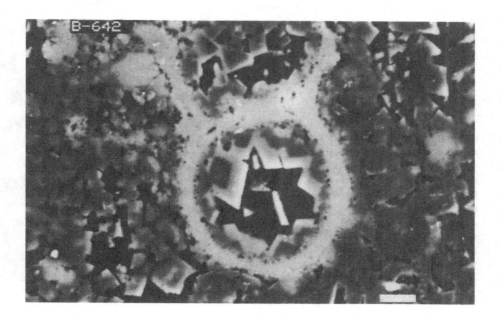

Figure 6.7. Complexly zoned rhombs within foraminifer test. The rhombs consist of a core of dolomite followed by calcite and ankerite. The zoning reflects precipitation in marine and mixing-zone fluids undergoing changing chemical composition. F = foraminifer test, c = calcite, d = dolomite, a = ankerite. The bottom photograph is an enlargement of the top photograph. Frio Formation (Oligocene), Texas Gulf Coast. Scale bars = 5 μm (top) and 2 μm (bottom). (From Brady and Krinsley, 1990, fig. 6, p. 5; reproduced by permission.)

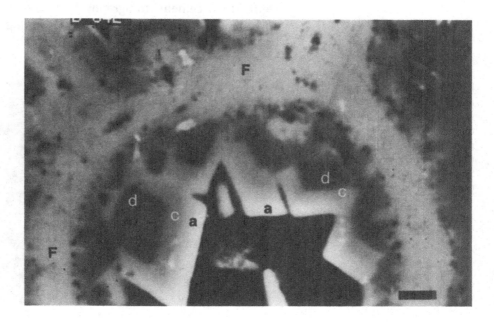

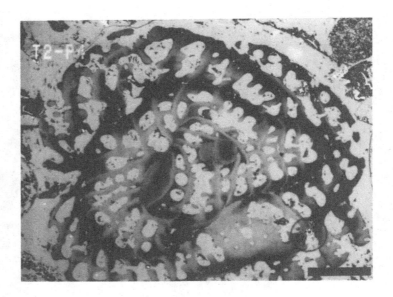

Figure 6.8. Foraminifer with macroscopic biomoldic porosity (black). The chambers of the foraminifer are nearly filled with low-magnesian calcite cement (white); undissolved shell walls appear gray. Holocene limestone, Schooner Cays, Bahamas. Scale bar = 200 μm. (From Budd and Hiatt, 1993, fig. 3F, p. 265; reproduced by permission.)

tated within a single pore. For example, a marine cement may be generated on the sea floor (Fig. 6.10), followed by a meteoric cement (if the sediment is subsequently uplifted into the meteoric zone), followed eventually by a burial cement precipitated after deep burial. With careful study it is possible to detect each of these generations of cement. This technique of studying successive generations of cement is referred to as **cement stratigraphy**. Various methods for studying cement stratigraphy are in use, including staining tech-

Figure 6.9. Calcite cement infilling algal connected tubules in a shallow marine calcirudite (biosparrudite). The bright dots are pyrite. Craighead Limestone (Upper Ordovician), Girvan, Scotland. Scale bar = 100 μm. (Photograph by Anton Kearsley.)

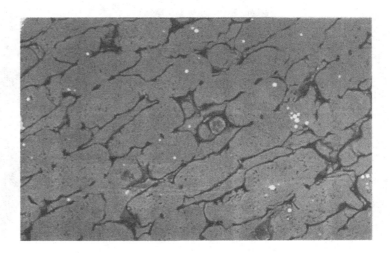

niques that reveal variations in ferrous iron content of the cement. The most common technique for studying cement stratigraphy, as well as the zoning in dolomites described above, is CL. CL in carbonates is caused mainly by the presence of trace elements, particularly Mn^{2+}, and to a lesser extent by other factors such as distorted crystal structures and compositional inhomogeneities. On the other hand, Fe^{2+}, in particular, tends to quench the luminescent reaction. Carbonate cements commonly exhibit a trend from oldest to youngest cements of (1) nonluminescence to (2) bright luminescence to (3) dull luminescence (Fig. 6.11). This trend has been interpreted to signify progressive change during burial from well-oxidized conditions that (1) inhibit the uptake of manganese and iron (no luminescence) to (2) reducing conditions that favor uptake of Mn^{2+} but keep Fe^{2+} locked up in interactions with organic matter and sulfates (bright luminescence) to (3) deeper-burial reducing conditions where Fe^{2+}

Figure 6.10. Marine, aragonite-needle cement precipitated within macropores in coral. Holocene carbonates, Manele Bay, Lanai, southern coastline, Hawaiian Islands. Scale bar = 100 μm.

is available to quench luminescence (dull luminescence). Thus, these trends are commonly regarded to be primarily due to relative concentrations of iron and manganese as a function of Eh and pH, although other investigators have suggested that additional factors (e.g., other trace elements) may also affect CL (e.g., Machel et al., 1991).

We are not aware that any investigators have used BSE to study cement stratigraphy; however, the possible applications are obvious. If successive generations of carbonate cements are characterized by differences in concentration of manganese and iron, as the available evidence suggests, BSE should reveal these differences as sharply as it reveals zoning in dolomites. Therefore, BSE appears to have the potential to become an important tool of the future for studying cement stratigraphy. In addition, the great magnification and enhanced resolution of BSE allow fine-scale details of compositional variations in cement to be easily detected. For example, Figure 6.12 shows zoned calcite cement in an ammonite chamber. The early-formed botryoid is composed of high-Mg calcite, with thin zones of Mn- and Fe-rich calcite. Figure 6.13 provides an additional example of carbonate cement within an ammonite chamber. In this example, variations in Fe/Mg content of the calcite give the cement a distinct mottled texture.

Figure 6.11. Schematic representation of banding in carbonate cements as revealed by cathodoluminescence. The variation in luminescence from nonluminescent through bright to dull reflects different generations of cements precipitated under different burial conditions. (From Choquette and James, 1987, fig. 21, p. 16; reproduced by permission.)

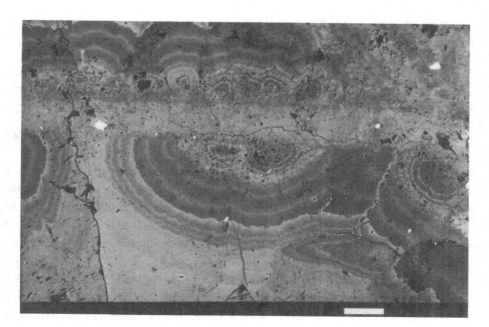

Figure 6.12 (above). Zoned calcite cement in an ammonite chamber. The early-formed botyroidal cement is composed of Mg-calcite, with thin zones of Mn- and Fe-rich calcite. Cleveland Ironstone Formation (Middle Lias), northeast England. Scale bar = 100 μm. (Photograph by Anton Kearsley.)

Figure 6.13 (below). Mottled carbonate cement in an ammonite chamber; mottling is caused by variation in Fe/Mg ratio in the cement. The septum edges are apatite-rich. The inner part of the shell also contains fine-grained pyrite (white). Cleveland Ironstone Formation (Middle Lias), northeast England. Scale bar = 100 μm. (Photograph by Anton Kearsley.)

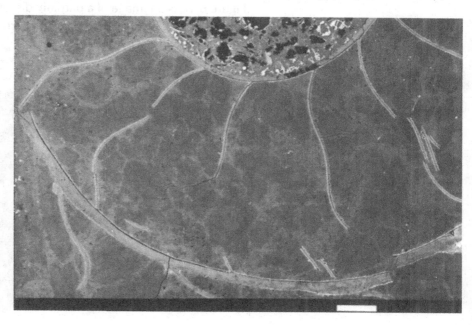

Internal Structure of Carbonate Grains and Other Structural Features

Because of its superior resolution, BSE provides a useful tool for studying the internal structure of fossils and other carbonate grains. If differences in chemical composition are present, details of the structure are revealed and different mineral phases (e.g., low-Mg calcite, high-Mg calcite, dolomite) are easily detected. For example, Figure 6.14 shows a foraminifer test composed of 4–5 μm crystals of low-Mg calcite. The fibrous character of the crystals is shown with remarkable clarity. Figure 6.15 is a cross-sectional view of a planktonic foraminifer that displays a dark central test area and a clear, bright outer cortex. The inner calcite layer of planktonic foraminifer tests is commonly deposited fairly rapidly and tends to incorporate some Sr and Mg into the calcite lattice. The outer cortex of calcite is deposited more slowly and generally consists of nearly pure calcite (W. N. Orr, personal communication). Figures 6.16 and 6.17 provide additional examples of the fine-scale detail of shell structure revealed in BSE images. Note again in Figure 6.15 the presence of a thick, clear outer cortex (nearly pure calcite) and a darker (Mg- and Sr-bearing) inner layer. Perforations in the shells (small dark spots) are also clearly revealed.

Figure 6.18 is an image of a partially dissolved foraminifer test that has been replaced in part by chert and iron oxide.

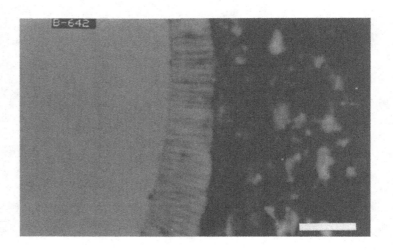

Figure 6.14. BSE image of a foraminifer composed of 4-5 μm crystals of low-Mg calcite along the edge. The center is composed of calcite cement. Frio Formation (Oligocene), Texas Gulf Coast. Scale bar = 5 μm. (From Brady and Krinsley, 1990, fig. 1, p. 3; reproduced by permission.)

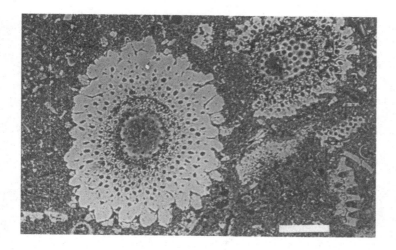

Figure 6.15. Cross-sectional view of a planktonic foraminifer that displays a dark, Sr/Mg-bearing, inner cortex and a thick outer cortex (bright) of nearly pure calcite. Middle Miocene nanofossil ooze, DSDP Leg 90, Site 594, off eastern South Island, New Zealand. 458 m below sea floor. Scale bar = 50 μm.

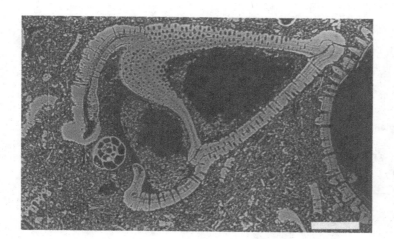

Figure 6.16. Cross-sectional view of a planktonic foraminifer (*Globorotalia*) test showing the fine-scale details of the shell structure revealed in a BSE image. The calcite outer cortex (bright) is nearly pure, and the Sr/Mg-bearing inner cortex is slightly darker. The small, dark spots are perforations for the pseudopodia. Middle Miocene nanofossil ooze, DSDP Leg 90, Site 594, off eastern South Island, New Zealand. 458 m below sea floor. Scale bar = 50 μm.

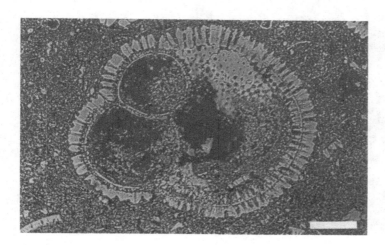

Figure 6.17. Cross-sectional view of a planktonic foraminifer (*Globogerina*) test showing the inner and outer cortex and shell perforations. Middle Miocene nanofossil ooze, DSDP Leg 90, Site 594, off eastern South Island, New Zealand. 458 m below sea floor. Scale bar = 50 μm.

Figure 6.18. A probable recrystallized foraminifer test, showing an
irregular exterior outline. The small
black areas within the test represent either microdissolution cavities or unfilled chambers (the large
black area is a plucked hole).

Bright specks within the test are
pyrite. Dolomite crystals (dark
gray) are scattered through the
calcite surrounding the test. Salem
Formation (Mississippian), Ozark
Dome, Missouri. Scale bar =
100 μm.

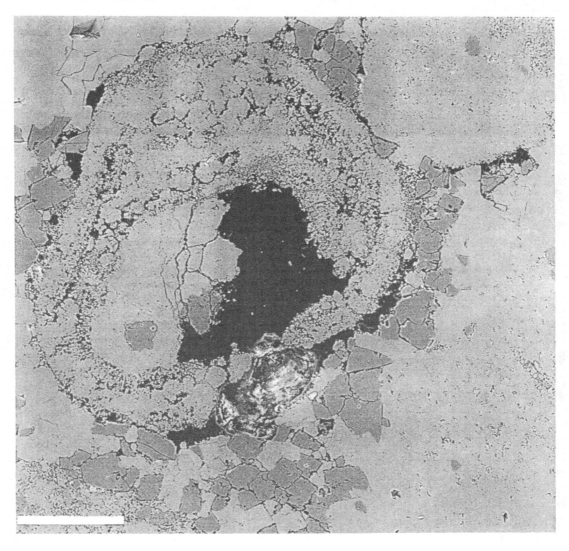

Fine-scale details, such as the microporosity visible in this photograph, would be impossible to detect by ordinary light microscopy. On the other hand, BSE is of little use for studying the interior details of fossils that are compositionally homogeneous. For example, a fossil composed entirely of nearly pure low-Mg calcite will display little or no internal structure in BSE.

Figure 6.19 is a BSE image of a partially dolomitized ooid. The individual dolomite laminae are dark in color and are engulfed by neomorphic calcite (light color). The fine resolution in the BSE image aids in interpreting the paragenetic sequence of diagenetic events in this ooid. It was composed initially of aragonite, which was partially dissolved with subsequent precipitation of dolomite prior to calcification of aragonite cortices (Zempolich and Baker, 1993). BSE images can help to identify many other kinds of diagenetic events. For example, Figure 6.20 shows microstylolites between two ooids, developed as a result of pressure solution. Dark bands within the ooids are rich in organic matter. By contrast, fracturing of ooids owing to compaction during diagenesis is illustrated in Figure 6.21.

Figure 6.19. Partially dolomitized aragonite ooid. Individual laminae (dark color) are 10–30 μm wide and are composed of coalesced euhedral to subhedral crystals 10–20 μm in size. Dolomite laminae are engulfed by neomorphic calcite (light color). These fabrics indicate that there was some dissolution of aragonite and precipitation of dolomite prior to calcitization of aragonite cortices. Beck Springs Dolomite (Proterozoic), Death Valley, California. Scale bar = 100 μm. (From Zempolich and Baker, 1993, fig. 5F, p. 601; reproduced by permission.)

Figure 6.20 (above). Partial dissolution of the outer cortex of an ooid (arrow) owing to pressure solution. The dark bands within the ooid cortex are rich in organic matter. Lincolnshire Limestone (Middle Jurassic), England. Scale bar = 100 μm.

Figure 6.21 (below). Carbonate ooid that displays well-developed, organic-rich cortex laminae. Part of the cortex has been broken and disrupted owing to burial compaction. Lincolnshire Limestone (Middle Jurassic), England. Scale bar = 100 μm.

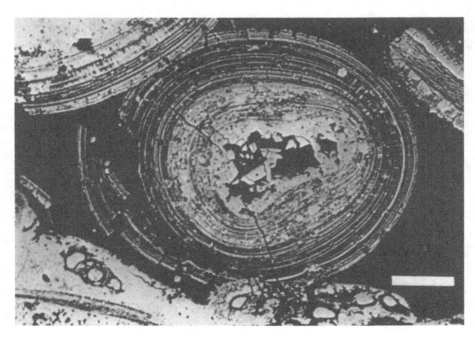

Carbonate Porosity and Microporosity

As in the case of sandstones, BSE images provide an important tool for studying the sizes and shapes of pores in carbonate rocks. Furthermore, as described in preceding and following chapters, automated techniques are available that allow quantitative evaluation of porosity. BSE methods are especially applicable in the study of microporosity in fine-grained carbonates (micrites) and within fossils and other carbonate grains. Much of the porosity in carbonate rocks is secondary porosity, created as a result of dissolution of carbonate grains or cements. For example, mouldic porosity created by selective dissolution of shell material is easily visible in Figure 6.22. Figure 6.23 illustrates three kinds of microporosity. Numerous irregular-shaped dissolution micropores are present in the lower left and right corners of the photograph. In the upper half, several larger, elongated,

Figure 6.22. Mouldic porosity (black) formed by preferential dissolution of shell material after cementation of the shells by calcite. Lincolnshire Limestone (Middle Jurassic), England. Scale bar = 10 μm.

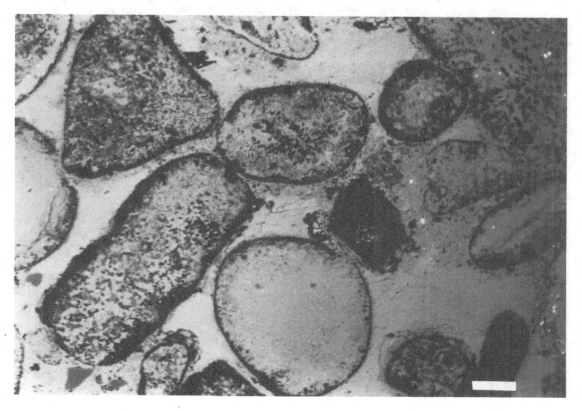

Figure 6.23. Secondary microporosity (black) in a limestone created as a result of random dissolution (lower right and left corners), dissolution along cleavage planes (top center), and fracturing (center and upper right corner). Salem Formation (Mississippian), Ozark Dome, Missouri. Scale bar = 10 μm.

oriented pores have been created as a result of dissolution along crystallographic planes in this carbonate crystal. The vertically elongated crystal in the lower, central part of the photographs displays porosity created by the presence of microfractures. Figures 6.8 and 6.15–6.17, which show considerable microporosity (black) within foraminifer tests, provide additional examples. Owing to differences in gray levels, BSE allows an operator to discriminate quickly between pore

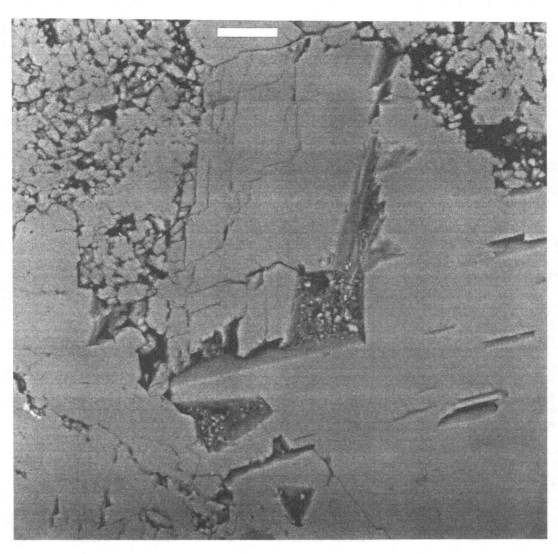

space and other dark areas in a grain that might be due to the presence of fine organic matter. The technique can be applied equally well to the study of intergranular porosity among carbonate grains, intercrystalline porosity in dolomites or micritic limestones, and fracture porosity.

Characteristics of Noncarbonate Minerals in Carbonate Rocks

BSE imagery can be used very effectively to study noncarbonate minerals that may be present in a carbonate rock. Atomic number contrast between carbonate minerals and noncarbonate minerals is normally quite large, causing such minerals to be clearly displayed in BSE images. Noncarbonate minerals may be present as detrital grains (e.g., quartz, feldspars, micas) or as diagenetic replacement minerals. Many kinds of minerals may replace carbonate minerals dur-

Figure 6.24. Barite (bright) replacing carbonate in a dolomitic limestone. Triassic, Durham, England. Scale bar = 100 μm.

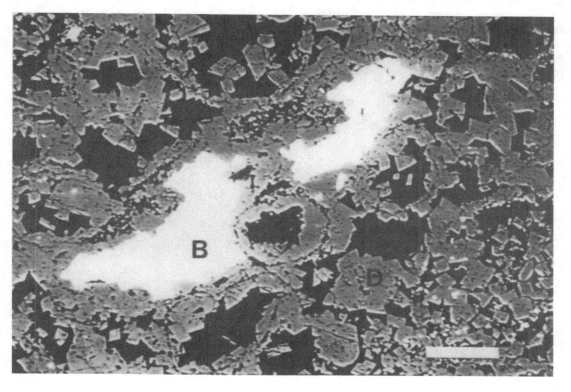

ing diagenesis, including microquartz (chert), pyrite, hematite, apatite, barite, and anhydrite. Figure 6.24 shows a barite grain (bright) that has replaced carbonate grains in a dolomitic limestone. Figure 6.25 shows a detrital mica, surrounded by fine carbonate minerals, in a dolocalcrete. Note that the mica has been split by displacive growth of calcite within the crystal.

Figure 6.25. Detrital mica grain (M) split by displacive growth of calcite (C). The carbonate cement surrounding the mica grain is mainly dolomite. Quaternary dolocalcrete, Kenya. Scale bar = 100 μm.

7

Desert (Rock) Varnish

INTRODUCTION

Desert (rock) varnish is a thin (< 100 to 500 μm thick) coating of ferromanganese oxides, clay minerals, and trace elements, which can form on sedimentary, igneous, or metamorphic rocks with stable surfaces. Microscopic study of varnish shows that it is characterized by many kinds of textures including numerous ultrathin laminae that commonly grade into irregular pods of jumbled material (Krinsley et al., 1990; Krinsley and Dorn, 1991). Thus, varnish is a special kind of microscale sedimentary deposit. It is most common on Quaternary-age rocks, but it has also been reported on Miocene, Triassic, and even Precambrian rocks. Both abiotic and biotic (mostly bacterial) processes may be involved in the formation of varnish.

Rock varnish is most common in arid environments (e.g., Krinsley and Dorn, 1991), but it also forms in more humid environments (e.g., Douglas, 1987). Because the origin of varnish may be related to climatic factors (Dorn, 1986), it is of particular interest to paleoclimatologists. It is also of interest to archeologists and geomorphologists, who have used the cation-ratio (CR) dating technique to determine the age of the varnish (Dorn and Whitley, 1984; Dorn et al., 1988; Pineda et al., 1990), which establishes a minimum age for a surface upon which varnish has formed. Geologists are especially interested in the origin and depositional history of rock varnish and the diagenetic changes that may affect it after initial formation.

To develop a better understanding of varnish origin, geologists are studying: (1) the uniformity of texture, chemical composition, and mineralogy of varnish on a micron scale; (2) the nature of the contact between varnish and the underlying rock surface, that is, whether it is sharp or gradational; and (3) textural evidence for the role of organisms in the formation and diagenesis of varnish.

One of the major problems investigators have faced in studying rock varnish is the small size of the particles present in varnish layers. Because the resolution of the light microscope (~2000 Å) is inadequate to reveal the nature of most of the particles in varnish, other methods of study are required. Prior to 1989, light microscopy and secondary scanning electron microscopy were the principal tools used to study microscopic detail in varnish. Krinsley and Anderson (1989) pioneered the use of backscattered electron microscopy (BSE) in the study of varnish, and they have shown that BSE is a far superior tool for discriminating the fine-scale features of varnish. Since this initial presentation, varnish research has turned increasingly toward BSE (e.g., Dorn and Dragovitch, 1990; Loendorf, 1991; Reneau et al., 1992; Carlos et al., 1993). Use of BSE to study the submicroscopic textural features and chemical composition of varnish has vastly increased our understanding of the characteristics and origin of varnish. In this chapter, we summarize some of the principal applications of BSE to the study and understanding of rock varnish.

DEPOSITIONAL TEXTURES AND STRUCTURES

Sedimentary Layers

Most rock varnish is characterized by the presence of layers ranging in thickness from several nanometers to a micron or more (Fig. 7.1; Krinsley et al., 1995; Krinsley, in press). The layers are believed to form initially by deposition of un-consolidated atmospheric dust. Alternations of light and dark BSE laminae are clearly revealed in the backscatter mode (Fig. 7.1). As much as 70% of the material in the dark col-ored layers may consist of clay minerals, mostly mixed-layer illite-smectite with some kaolinite. Fe and Mn minerals, to-gether with minor amounts of micas, feldspars, and quartz, make up the remainder (Potter and Rossman, 1977, 1979; Krinsley et al., 1990). The lighter layers contain a higher pro-portion of Fe and Mn minerals, and both Fe-rich and Mn-rich laminae (identified by X-ray analysis, or EDX, in conjunction with BSE) may be present. This alternation of chemically dis-tinct layers is due, at least partially, to diagenetic processes such as solution and precipitation following deposition, in-cluding action by Mn- and Fe-fixing bacteria. The layers may be discontinuous over micron or submicron distances.

Figure 7.1. Layered varnish from Marie Byrd Land, Antarctica, from a slope off Mt. Van Valkenburg, Clark Mountains. Well-developed varnish layers extend from the varnish surface to about 40 μm below the surface. The time at which varnish formation was initiated is unknown. Scale bar = 100 μm.

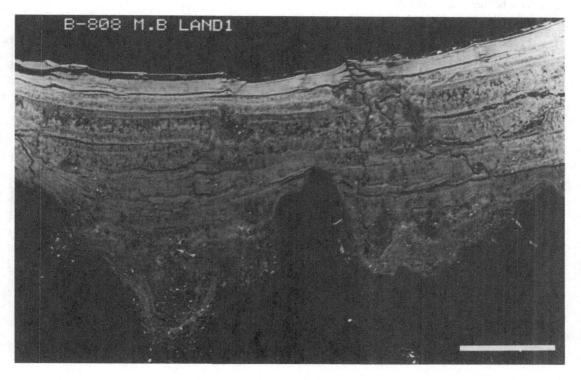

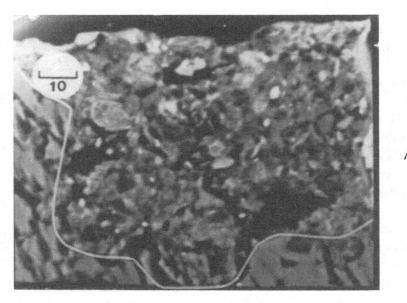

A

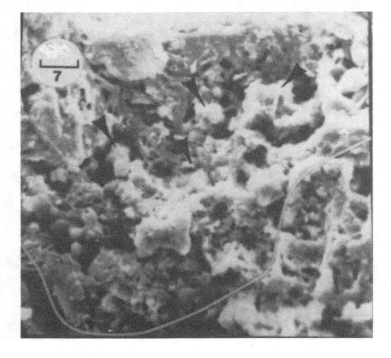

B

Figure 7.2. Depression in a rock surface filled with a nonlayered, chaotic mixture of typical varnish constituents (manganese/iron oxides and clays) and a high concentration of aeolian detritus such as the barium sulfate crystals identified by arrows in (B). (A) A BSE photograph of varnish from the Marble Mountains, Mojave Desert, California (Late Pleistocene). (B) An SE photograph of varnish from Crater Flat, southern Nevada (Late Pleistocene). Age of the Crater Flat sample is about 20,000 years, as determined from soil development and radiocarbon and uranium-series measurements. Rock/varnish contacts are indicated by white lines. Scale bars: (A) = 10 μm; (B) = 7 μm.

Pits

Pits or depressions in rock surfaces where eolian dust tends to accumulate may be present in varnish (Fig. 7.2). These depressions may be the result of erosion by microcolonial fungi or other acid-secreting organisms, chemical etching, or simply natural irregularities. They are very small, on the order of microns in diameter.

The pits may contain chaotic mixtures of various minerals (Fig 7.2), which can be identified by using EDX in conjunction with secondary electron (SE) and BSE imaging. Typical minerals consist of Mn and Fe oxides of diagenetic origin, eolian detritus such as quartz and barite, and clay minerals, which may be partly diagenetic minerals and partly eolian debris. As these minerals accumulate, the varnish apparently aggregates fairly rapidly in the pits, as opposed to slower rates of growth in well-layered varnish (see Fig 7.1).

Some discontinuities in layers are created when a previously formed, layered varnish is subsequently leached by small amounts of capillary water that originate from desert dew (Glennie, 1970, pp. 20, 190) or infrequent desert storms. These layers may cut across previously formed ones. In some varnish, the chaotic portion may grade into layered varnish. It is difficult, however, to distinguish between chaotic textures that are the result of deposition and those that are the result of leaching and drying, which may form pores such as those shown in Figure 7.1, unless cross-cutting relationships are present.

Evidence of Erosional Events

Once a section of varnish has formed, subsequently it may be partially eroded by eolian action, followed by deposition of a second layered varnish. Successions of such erosional and depositional events can be observed in BSE or SE photographs, as shown in Figure 7.3. Thus, it may be possible to interpret a series of ancient climatic events by studying microscale erosional and depositional events within specific varnishes. Numerical dating of varnish (Dorn et al., 1992) might even allow the ages of these events to be estimated, if it becomes possible to date individual layers within varnish.

DIAGENESIS

Fe and Mn Concentrations

The process or processes by which Fe and Mn are concentrated in desert varnish are still poorly understood. Mn concentrations in black varnish may be as much as 50 times higher than those in the underlying rock (Dorn and Oberlander, 1982; Jones, 1991). For example, bulk analyses of Mn and Fe concentrations in varnish from the Mojave Desert typically show ranges of 10–15% for both elements. Although abiotic processes may be involved in the concentration process, many researchers (e.g., Dorn and Oberlander, 1981; Palmer et al., 1985; Jones, 1991; Drake et al., 1993) believe that bacteria or other microorganisms may be mainly responsible. Bacteria are known to concentrate Mn as rods, cocci, and filaments, as shown in Figure 7.4. Some micron-

Figure 7.3. A succession of depositional and erosional varnish events is recorded in this SE photograph of varnish on a 2-m ventifacted boulder from the Mojave Desert, California. Three phases of varnish buildup and two to three phases of varnish erosion are indicated: (1) the lower varnish layer was deposited; (2) the area beneath the double arrow was eroded, probably owing to secretion of organic acids by fungi or lichens; (3) varnish filled the resulting depression; (4) eolian abrasion truncated the varnish above the double arrow; (5) a new layer of varnish was deposited above the double arrow; and (6) some eolian abrasion may have occurred on top of the newest varnish layer. Scale bar = 4 μm.

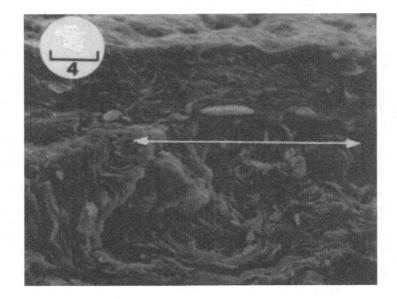

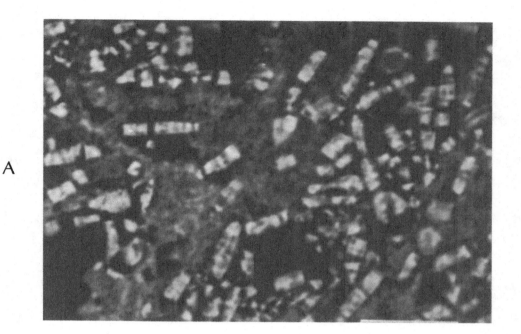

A

Figure 7.4. Bacteria associated with varnish formation. (A) Varnish from the surface of a quartzite boulder, Hanaupah Canyon alluvial fan, Death Valley, California, about 50,000 years old. The approximately 1-μm-wide bright spots in this picture may be fossil bacteria encapsulated in manganese oxides. (B) Probable bacterial casts forming rods elongated in bedding direction (arrows) in porous texture of varnish from a petroglyph-marked rock surface, south Australia. Radiocarbon age of the varnish is > 40,000 years. Electron-probe measurements indicate a varnish composition of > 7% iron and manganese oxides. Scale bars: (A) = 10 μm; (B) = 20 μm.

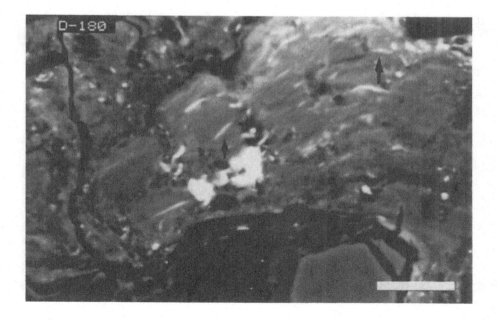

B

A

Figure 7.5. Stromatolite-like structures in varnish. (A) Varnish on a quartzite boulder, Death Valley alluvial fan (late Pleistocene), Death Valley, California. Note the roughly parallel layering (arrows) near the rock-varnish contact and the stromatolite-like layering within the varnish. Most of the very bright layers are high in Mn; the black layers are open cracks. (B) Varnish, displaying stromatolite-like layering, from the eroded surface of a fumerole mound on the Bishop Tuff (734,000 years old). The black microdepressions (white arrows) may have been eroded by acids secreted by microcolonial fungi and subsequently filled by quartz, feldspar, and barium sulfate. The black arrow points to a carbon artifact left during specimen coating. Scale bars: (A) = 100 μm; (B) = 10 μm.

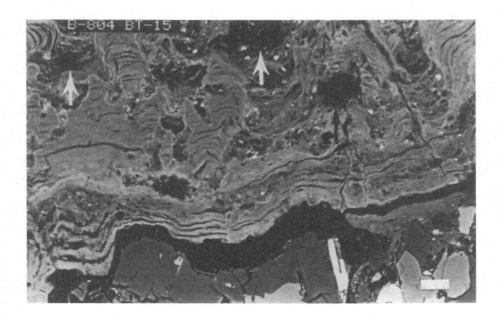

B

scale deposits of Mn-rich material may be fossilized bacter-ial casts (Fig. 7.4A, B; Krinsley, in press). Stromatolite-like features in varnish (Fig. 7.5A, B) have also been attributed to biological growth processes (Krinsley et al., 1990).

When initially formed, young varnishes are soft (Dorn and Oberlander, 1982). Varnish begins to harden with age, but may be softened again by water percolating through the varnish from above or below. Thus, numerous episodes of wetting (desert storms and dew) and drying (high daytime temperatures) may occur, which give rise to various deforma-tion features. The most striking of these features are mi-croscale, low-angle folds (anticlines and synclines), proba-bly produced by the pressure of wetting and drying (see Fig. 7.1). These physical deformation processes may be accom-panied by fluid movement, resulting in precipitation of Mn and Fe oxides and formation of discontinuous laminae.

If varnish becomes sufficiently cohesive (hardened) dur-ing wetting and drying, deformation pressure can produce microfractures (Fig. 7.6). Fractures oriented both parallel, perpendicular, and at an angle to the bedding are common. The fractures may be unfilled or partially or completely filled, mostly with Mn/Fe oxides. The filling material com-monly includes Si, Al, and other elements characteristic of clay minerals. Very thin feeder veins containing these materials may branch off from larger filled fractures. So-called fumerole mounds sit on top of some fractures that penetrate entire varnish sections (Fig. 7.6, arrow). Fractures that end in mounds commonly have high concentrations of Mn and Fe. The fractures apparently facilitate the movement of Mn- and Fe-rich fluids, allowing the Mn and Fe oxides to precipitate on the interface between the atmosphere and var-nish.

Varnish that occurs in depressions may display distinct differences in Mn/Fe concentration from that in surrounding varnish layers (Fig. 7.7). Varnish in depressions commonly consists of dark layers (BSE images) with irregular holes, which are oriented with their long axes roughly along the

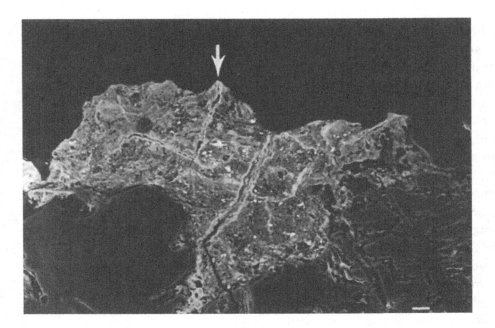

Figure 7.6. Irregularly eroded varnish on the surface of a quartzite boulder, Starvation Canyon alluvial fan (late Pleistocene, 10,000–50,000 years BP), Bishop Creek, eastern California. The varnish is almost completely eroded from the rock surface on the right side of the photograph. The varnish displays fractures partially filled with iron and manganese oxides (very white material). The very bright dots are probably iron and/or manganese oxides and barium sulfate. Scale bar = 10 μm.

Figure 7.7. Partially leached varnish on a graniodiorite boulder from a glacial moraine (140,000 years BP), Bishop Creek, Death Valley, California. Mg-poor black varnish (right side) has been leached, probably by running water. At the top of photograph, from right to left, the color changes from black to dark gray to lighter gray, indicating increase in manganese and iron concentration from right to left. Scale bar = 10 μm.

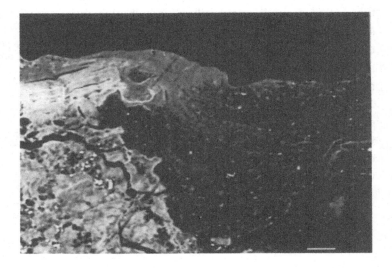

layers. These dark layers (e.g., right portion of Fig. 7.7) contain low concentrations of Mn and Fe. They commonly occur adjacent to normal, bright varnish (left portion of Fig. 7.7). Gradation from dark, to gray, to bright varnish may be present (top portion of Fig. 7.7). The gray layers contain more Mn and Fe than do the dark layers, and the bright layers are normal varnish with the typical concentration of Mn and Fe.

Variations in Mn and Fe concentration of this kind are probably the result of leaching. Leaching takes place during wetting and drying, perhaps during day/night cycles of desert dew. Mn and Fe that accumulate in normal varnish layers are removed by water that slowly circulates through holes and cracks in the varnish. The average thickness of varnish is less than a few hundred microns; therefore, very little water is needed to fill available pore space. If water continues to move through the pores, creating a dynamic interchange of ions, severe leaching of Mn and Fe can occur.

DISCUSSION

The BSE photographs presented in this chapter illustrate that BSE can be an extremely powerful tool for the study of microscale textures in sedimentary materials, particularly when mineralogy and chemistry of the microfeatures observed must be established. In the case of desert varnish, BSE imaging provides clues to the physical, biological, and chemical processes that form varnish and the source of its constituents. The new technique, in conjunction with EDX, suggests that there is a great deal of fine variation in all these textures at the micron and even nanometer scales (Dorn and Oberlander, 1982; Dorn, 1984; Krinsley et al., 1995; Krinsley, in press). During the past 15 years, significant advances have been made in the development of experimental dating of varnish, including microlaminations, tephrachronology, uranium series, and other approaches (see Dorn, 1994). Yet, until varnish formation and diagenesis are better understood, all these approaches must be viewed as experimental. BSE has played a critical role in exposing uncertainties associated with dating varnish (Krinsley et al., 1990), and it continues to play a key role in the development of various models of varnish development.

8

Glauconite

INTRODUCTION

The term "glauconite" applies loosely to a group of green, three-layer clay minerals, all of which are chemically complex K-Al-Fe silicates containing more than about 15% total Fe_2O_3. Glauconites range in composition from K-poor smectites to K-rich glauconitic micas, with a general trend of increasing K with increasing age. These minerals commonly occur in sediments as small, rounded grains or peloids. Glauconite grains are particularly abundant in Middle Cambrian to Early Ordovician and Middle Cretaceous to Early Cenozoic rocks (Van Houten and Purucker, 1984). They are common also in many modern environments, such as the continental shelf off Vancouver Island, British Columbia, Canada; Queen Charlotte Sound, British Columbia; Monterey Bay, California; the Atlantic coastal shelf off the United States; the East Australian continental margin; and in bottom sediments in various parts of the Atlantic, Pacific, and Indian oceans.

The origin of glauconite is not completely understood. Most workers believe that it forms authigenically on the sea floor by alteration of substrate materials such as skeletal debris, fecal pellets, and various kinds of mineral grains, particularly mica and feldspars (e.g., Boggs, 1992, p. 152; Chaudhuri et al., 1994). Some glauconites may have formed by

alteration or transformation of mixed-layer clays by adsorp-
tion of K and Fe. Others likely form by precipitation of
dissolved material in the pores of a substrate that is progressively altered and replaced (Odin and Matter, 1981). Precipitation can take place also in the cavities of microfossils to
produce internal molds. Berner (1971) reported that glauconite forms slowly at the sediment/water interface, where it
is associated with organic matter under generally positive,
but fluctuating, Eh conditions. Detailed studies of sediments
from the East Australian continental margin show that glauconite is forming today at sites located at the transition between relict Fe-oxyhydroxide-rich, organic-poor outer-shelf
sediments and fairly rapidly accumulating Fe-oxyhydroxide-
poor, organic-rich deeper-water sediments (O'Brien et al.,
1990). Unlike most workers, Lipkina (1990) suggested that
glauconite in sedimentary rocks and sediments is not an authigenic mineral but, rather, is a hydrothermal green clay
that is redeposited in sediments as lumps.

Organisms and organic matter may be involved directly or
indirectly in the formation of glauconite. Organic matter
supposedly provides the appropriate redox potential to initi-
ate the process (Odom, 1984; Chaudhuri et al., 1994); however, evidence presented by Chaudhuri et al., indicates that
the presence of organic matter may not be required in all
cases. For example, minute blebs of glauconite are present
within detrital feldspar grains in Proterozoic sandstones of
South India, aligned along cleavage planes or filling dissolution cavities. The presence of glauconite within the feldspars,
which do not contain organic matter, suggests that organic
matter is not necessary for glauconite formation, although
organic matter may influence the redox potential of associated pore waters.

Whatever its exact mode of origin, glauconite appears to
form preferentially in shallow-marine environments. Thus,
the presence of glauconite in ancient sedimentary rocks is
regarded as a fairly reliable indicator of marine, dominantly
shelf environments. On the other hand, glauconite has been

reported, on the basis of stable isotope data (Xie and Shen, 1991), to form also in some nonmarine environments, although nonmarine glauconite is relatively uncommon. In addition to its importance as an environmental indicator, it is the most useful mineral for direct age determination of sediments by use of the $^{40}K/^{40}Ar$ dating method (e.g., Odin and Dodson, 1982).

Using backscattered electron microscopy (BSE) to study glauconite textures has the potential to provide additional insight into the environmental significance of glauconite as well as its reliability in age determination. Compared to the light microscope, BSE enables observation of smaller details both within and around glauconite grains, yielding more detailed insights into processes of formation. In addition, observations of chemical and physical variations within individual glauconite grains by utilizing BSE can provide information that may be useful for refining the $^{40}K/^{40}Ar$ dating method. For example, the presence of cracks within glauconite grains may indicate that the grains are not completely closed systems. Also, the presence of nuclei grains, such as feldspars, that contain ^{40}K could yield incorrect ages if such grains are used in age determination.

GLAUCONITE TEXTURES REVEALED BY BSE

Fissured Grains

Glauconite grains can have a variety of shapes, including spheroidal, ovoid, botryoidal, tabular, and vermiform. They occur also as internal molds and casts of microfossils. Modern glauconite peloids display both smooth, unfissured grains and deeply fissured or cracked grains (e.g., Bornhold and Giresse, 1985). Fissured grains are also common in older sediments. Cracks in modern and Quaternary to late Tertiary glauconite peloids are commonly open (Figs. 8.1 and 8.2). Fractures in older glauconite grains tend to be filled with later generations of glauconite, with a composition slightly

Figure 8.1 (above). Glauconite grains displaying well-developed cracks, some of which have been filled by a later generation of glauconite. Some of the glauconite grains are darker than others, indicating slight differences in chemical composition. A few foraminifers (f) are also present. Core photograph, Tasman Sea off the western coast of Tasmania. Late Tertiary. Scale bar = 200 μm.

Figure 8.2 (below). Close-up photograph of a portion of Figure 8.1. Second-generation glauconite (dark) fills cracks, as in the grain in the lower left corner. Some cracks are unfilled and cut across second-generation glauconite (arrows). Glauconite grains in this photograph display three different gray-scale intensities, suggesting multiple episodes of glauconite formation or alteration. Core photograph, Tasman Sea off the western coast of Tasmania. Late Tertiary. Scale bar = 100 μm.

different from that of the original grain (Fig. 8.3), or with matrix minerals such as clay minerals. The presence of such filled cracks demonstrates that the cracks themselves are not artifacts of sample preparation. The origin of cracks in glauconite peloids is not fully understood. They have been attributed to shrinkage owing to loss of interlayered water (Odom, 1976) and to displacive growth of glauconitic minerals in initially smaller substrates (Odin and Matter, 1981).

Figure 8.4 shows several fissured, Tertiary glauconite grains that display slight modifications of their original chemical composition (less K and Fe in the darker areas) along their edges. The cracks may have provided access by fluids to bring about the chemical changes, although some cracking probably occurred after the latest episode of chemical alteration. Some glauconite grains display internal structures and compositional differences that indicate a complex history of crack formation, crack filling, and chemical alteration. The large glauconite grain in Figure 8.5, for example, has undergone at least five stages of development: (1) initial formation of a grain having the composition of the light areas, (2) cracking of the grain, (3) infilling of the cracks and replacement of portions of the original grain with Fe/K-poor glau-

Figure 8.3. Large, fissured glauconite grain (center of photograph), with secondary glauconite in cracks. Smaller glauconite grains and carbonate fossils (foraminifers) are also present. The surrounding matrix consists mainly of glauconite, which contains less iron than the glauconite grains. This sediment formed in shallow water with low sedimentation rates. Eocene-Oligocene, Southwest Rockall Plateau, North Atlantic Ocean. Scale bar = 100 μm.

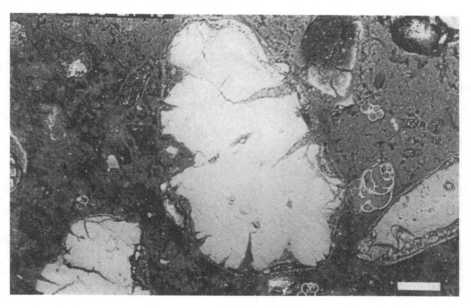

Figure 8.4. Irregular-shaped glauconite grain with unfilled cracks. Different gray-scale intensities indicate that at least two generations of glauconite are present. The cracks penetrate both generations of glauconite, suggesting that much of the cracking occurred after the second-generation glauconite formed. Core photograph, Tasman Sea off the western coast of Tasmania. Late Tertiary. Scale bar = 50 μm.

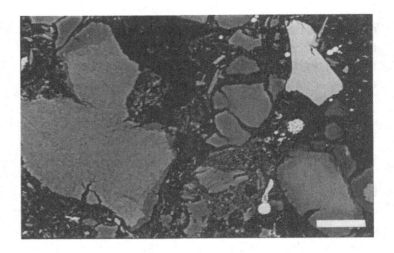

Figure 8.5 (below). Several generations of glauconite are present in this large glauconite grain, which displays a succession of cracks and fillings. Portions of the outside of the grain are filled with matrix material, which grades into glauconite. Formed in shallow water during a marine transgression under conditions of very slow sedimentation. Eocene-Oligocene, Southwest Rockall Plateau, North Atlantic Ocean. Scale bar = 50 μm.

conite (dark areas), (4) a second episode of crack formation, producing the small fissures, (5) filling of some of these small cracks by Fe/K-poor glauconite. The complex history of this grain suggests that some glauconite grains may not become closed systems with respect to gain or loss of K or A until quite late in their development – a factor of importance in assessing the reliability of glauconite for the purpose of age determination. Such detailed history of a glauconite grain could not be determined by petrographic microscopy.

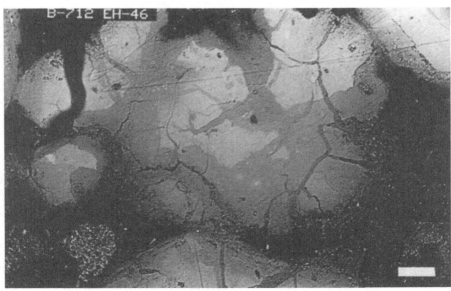

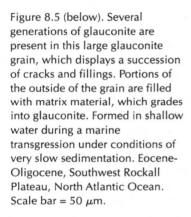

Replaced and Infilled Fossils

Glauconite peloids may engulf and replace fossils as they grow within a substrate. For example, Figure 8.6 shows a large glauconite grain that has replaced all or parts of several foraminifers. Some incompletely replaced fossils show up as "ghosts" within the glauconite grain (arrow). This glauconite grain is clearly authigenic; the area now occupied by this grain was originally composed of fossils and matrix material. Figure 8.7 provides an additional example of fossil replacement by glauconite. Some of the small planktonic foraminifers have been partially replaced by glauconite or infilled with glauconite cement; the larger glauconite grains could be replaced pellets or superficial ooids rather than fossils. Figure 8.8 shows a magnified (480 times) portion of the rim of the large glauconite grain in Figure 8.7. This rim appears to be the remnant of an ooid or pellet surface, consisting of layers of broken calcite (CA) infilled and surrounded by introduced glauconite (GL). The small, North–South oriented cracks in the large grain are filled with mixtures of glauconite and calcite.

Figure 8.9 is a glauconite grain that contains two fossil

Figure 8.6. Glauconite grain replacing matrix and carbonate fossils. The fossil ghost (arrow) is located within the large center grain. The glauconite is formed in deep water within a nanofossil-foraminifer ooze lying unconformably above Eocene-Oligocene sediments. Southwest Rockall Plateau, North Atlantic Ocean, Scale bar = 100 μm.

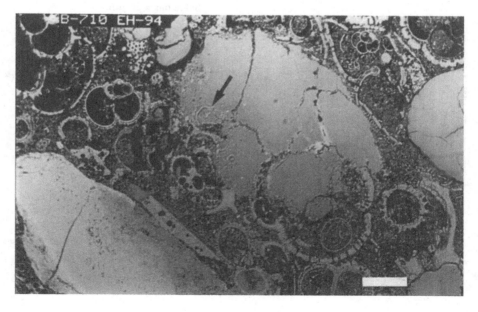

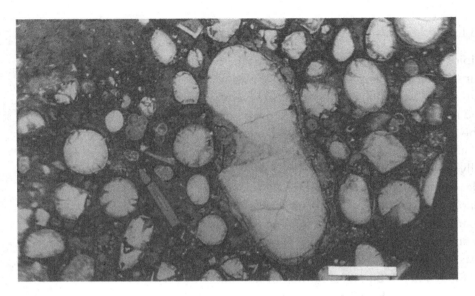

Figure 8.7 (above). Large glauconite peloid (center) surrounded by rounded and semirounded, chemically uniform glauconite peloids that may replace fossils or coated grains. Small foraminifers are scattered throughout the clay matrix. Eocene-Oligocene, Southwest Rockall Plateau, North Atlantic Ocean. Scale bar = 500 μm.

Figure 8.8 (below). Close-up image of the large grain shown in Figure 8.7. The inner part of the grain is chemically uniform glauconite, marked by the presence of two small cracks. The edge is composed of thin, distinct layers of glauconite (GL) and calcium carbonate (CA), but small grains of each appear within the other. Eocene-Oligocene, Southwest Rockall Plateau, North Atlantic Ocean. Scale bar = 20 μm.

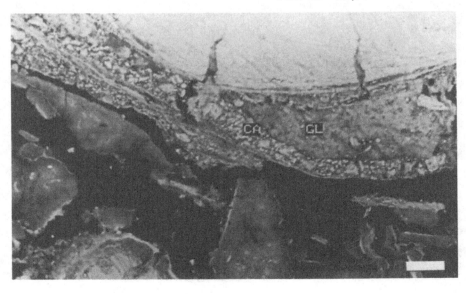

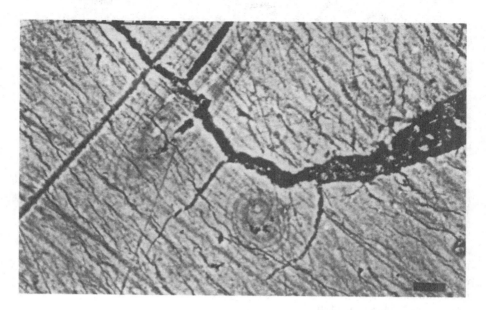

Figure 8.9. Highly magnified portion of a glauconite grain containing two fossil foraminifer ghosts. Eocene-Oligocene, Southwest Rockall Plateau, North Atlantic Ocean. Scale bar = 10 μm.

ghosts. These ghosts show up because they contain slightly more Mg than their surroundings and thus are darker. The grain is penetrated by several small and one, partially filled, large cracks. The variation in BSE contrast on a micron scale throughout this grain indicates a high degree of chemical variability. At present, BSE analysis is the only way that these small-scale features of glauconite grains can be observed and evaluated.

Figure 8.10 shows a fragment of a pelecypod (?) shell, with partially preserved original shell structure, which has undergone patchy replacement by other minerals (dark areas). A closeup of part of this bioclast, magnified 2000 times, is shown in Figure 8.11. EDX analyses within the dark areas indicate a glauconite chemical signature; however, the BSE contrast shows that other minerals are also present. Although BSE resolution allows mineral phases as small as 1 μm to be observed, such small areas cannot be separately analyzed by EDX, which tends to give an averaged chemical composition for grains of such small size.

Figure 8.10 (above). Fossil carbonate shell (pelecypod (?)), filled with patchy glauconite. Impure glauconite surrounds the shell. Eocene-Oligocene, Southwest Rockall Plateau, North Atlantic Ocean. Scale bar = 10 μm.

Figure 8.11 (below). Closeup image of Figure 8.10. The patchy glauconite within the shell is chemically complex. Although tiny crystals of other minerals are also present, the chemical signature obtained by X-ray analysis (EDX) is approximately that of glauconite. Eocene-Oligocene, Southwest Rockall Plateau, North Atlantic Ocean. Scale bar = 10 μm.

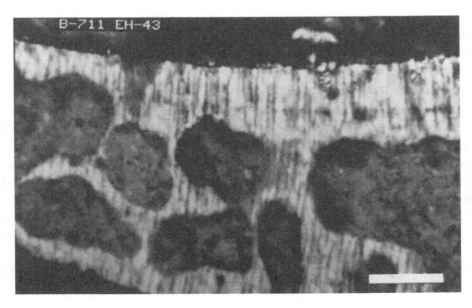

Glauconite Grains with Nuclei

Some glauconite peloids have a nucleus consisting of a large mica flake, quartz or feldspar grain, or other mineral. These grains apparently acted as nucleation centers around which the glauconite peloid developed. In some peloids, the growing glauconite developed as concentric layers (Fig. 8.12) similar to those of carbonate ooids. The nucleus in Figure 8.12 appears to be a rock fragment of some kind, which has been partially altered to glauconite that subsequently cracked. The cracks are filled with younger glauconite. Also evident is minor cracking of the glauconite cortex and in-filling with a younger glauconite cement. Most glauconite peloids do not display internal concentric layers in BSE images (Figs. 8.13–8.16). The nuclei may be single mineral grains of various sizes (Fig. 8.13) or multiple mineral grains (Figs. 8.14 and 8.15). Although some nuclei may interact chemically with the glauconite peloids as they grow (Fig. 8.16), most seem to have acted simply as centers for nucleation and growth.

The nucleus in Figure 8.13 is a calcium plagioclase grain that has undergone partial albitization along cleavage or twin planes, producing a pattern of parallel dark gray streaks

Figure 8.12. Large, concentrically layered glauconite peloid that encloses a complex nucleus containing filled cracks. The nucleus is composed of patches of glauconite with a different chemical composition than the outer layers of the peloid. Eocene-Oligocene, Southwest Rockall Plateau, North Atlantic Ocean. Scale bar = 100 μm.

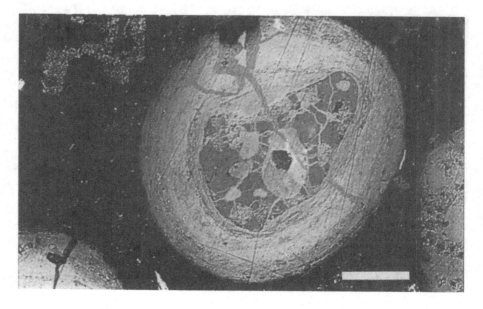

Figure 8.13 (above). Glauconite peloid with a large, calcic plagioclase feldspar nucleus. The plagioclase grain has been partially altered to albite along cleavage or twin planes. The matrix surrounding the nucleus contains quartz, carbonate, iron oxide, illite, micas, and feldspars. Whitby Mudstone Formation, Lias (Lower Jurassic), Cleveland Basin, U.K. Scale bar = 20 μm.

Figure 8.14 (below). Elongated glauconite peloid (center) containing two quartz grains as nuclei. The surrounding, oriented matrix contains quartz, carbonate, iron oxide, clay minerals (including kaolinite, illite, and chlorite), feldspars, and pyrite. Whitby Mudstone Formation, Lias (Lower Jurassic), Cleveland Basin, U.K. Scale bar = 50 μm.

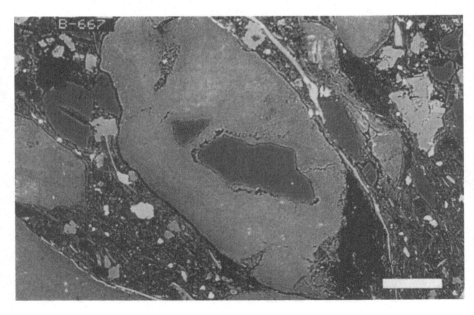

within the plagioclase nucleus. The very thin, bright layer on the left of the feldspar grain is late-stage chlorite. The peloid is surrounded by a matrix consisting of quartz, carbonate, iron oxide, illite, micas, and feldspars; nonmatrix grains of minerals such as quartz and carbonate are also present.

Figure 8.14 shows an irregularly ovoid glauconite peloid with quartz nuclei. The small quartz grain originally may have been attached to the larger but was subsequently separated along a microfracture by solution and growth of glauconite between the grains. Note that the glauconite peloid has been cracked (and filled) in a few places and that the lower right end (black area) has been dissolved. Growth of the large peloid caused compression of the surrounding matrix, for example, along the left middle part of the peloid. Late-stage chlorite (thin, bright line) is present along the upper right edge of the grain.

The glauconite peloid in Figure 8.15 has developed around several small quartz grains and altered illitic, mica grains. Both the quartz and mica grains appear to have been chemically etched along edges and microfractures. Some of this dissolution may have occurred during growth of the glauconite peloid; subsequent access by pore fluids along cracks likely caused additional etching. The matrix material surrounding the peloid is much the same as that in Figure 8.14.

Figure 8.15. Rounded glauconite peloid (center) containing nuclei of small quartz grains and altered illitic mica. Tiny blebs of pyrite (bright) are present in the surrounding matrix, which displays crude orientation. Whitby Mudstone Formation, Lias (Lower Jurassic), Cleveland Basin, U.K. Scale bar = 20 μm.

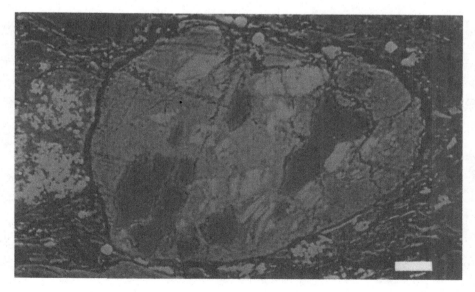

Figure 8.16 shows a glauconite peloid with an indistinct nucleus consisting of altered mica, which grades almost imperceptibly into the surrounding glauconite. Chemical elements leached from the mica nucleus appear to have been redistributed around the nucleus, mainly within the upper right portion of the glauconite grain. Small (bright) mical flakes in the left side of the glauconite grain may have been incorporated into the growing glauconite peloid as it partially engulfed the surrounding matrix.

Figure 8.16. Portion of a glauconite peloid with an internal fibrous nucleus, which may be altered mica that has reacted chemically with the surrounding glauconite. Whitby Mudstone Formation, Lias (Lower Jurassic), Cleveland Basin, U.K. Scale bar = 10 μm.

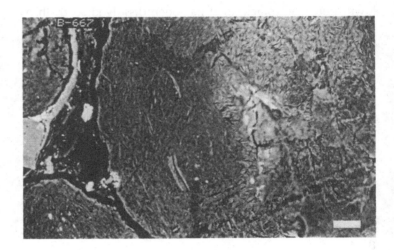

9

Image Analysis

INTRODUCTION

Image analysis is defined as the technique of obtaining quantitative information by measuring the features in an image (Mainwaring and Petruk, 1989). Image analysis has wide application in all fields of science today, including medicine, biology, physical metallurgy, remote sensing, military detection, food sciences, and the earth sciences. It has assumed particular importance in the mineral sciences owing to the need to acquire large amounts of quantitative chemical/mineralogical information together with morphological data. These requirements make application of image analysis in the mineral sciences more complex, sophisticated, and demanding in terms of instrumentation and automation than is the case for most other sciences.

Although the human eye is good at recognizing patterns and spatial relationships, it cannot make quantitative measurements from an image. Image analysis, on the other hand, has the ability to acquire and process huge amounts of quantitative data. Thus, image analysis lends itself to a range of applications in the fields of sedimentology and sedimentary petrology that include quantitative mineralogical and chemical analysis; analysis of grain size; particle shape and roughness; orientation of fabric or structure; pore size, shape, and volume; and the nature of boundaries between grains.

Images required for analysis can be produced by a variety of instruments, including petrographic microscopes, transmission electron microscopes, and scanning electron microscopes (SEMs). Images generated by SEMs include backscattered electron microscopy (BSE) images, secondary electron (SE) images, cathodoluminescence (CL) images, and X-ray element maps. BSE images are particularly useful for studying features in sedimentary rocks, as described in Chapters 3–8 of this book. Therefore, the techniques of image analysis discussed in this chapter focus on use of backscattered images. Because image analysis is a technique for characterizing, classifying, and comparing images by using numerical values for properties of features in images, application of image analysis requires identification and measurement of these features. The main feature properties are length, width, area, boundary irregularities, roundness, sphericity, size distributions, distances between features, clustering of features, total number of features, number of features that have specific properties, and textural relationships between features, including particle orientation (Petruk, 1989a). A researcher may make individual measurements for each feature in the image (feature specific) and/or global measurements for all features in the image (field-specific). Therefore, all the BSE images presented in this book can be used in some aspect of image analysis. BSE images to be used for image analysis may be acquired by photographic recording or by direct interfacing from an SEM to a digital frame store.

This chapter describes the fundamental techniques of image processing and analysis, which include five basic steps or stages: (1) identification of the task, which has a bearing on the pixel resolution; (2) image acquisition – direct digital acquisition or acquisition through an intermediate photograph; (3) image processing to enhance image quality; (4) image analysis; and (5) interpretation.

IDENTIFICATION OF THE TASK

A variety of feature properties, such as length, width, boundary irregularities, and textural relationships between features, are present in BSE photographs or digital images. The task of image analysis is to identify and measure those features that are important to a particular research objective. In the earth sciences, image analysis can be used for a variety of tasks such as automated, quantitative mineral identification; characterization of heavy mineral separates; automated measurement of mineral matter in coal; acquisition of chemical data; measurement of grain size and/or shape; quantitative measurement of porosity and pore shapes in sediments or soils; study of grain orientation or other oriented fabrics or structures; and integration of geoscientific map data (e.g., Petruk, 1989c). Once the research objective has been identified, commercial image analysis computer packages are available that can be tailored to meet the needs of the specific task at hand.

ACQUISITION OF DIGITAL IMAGES

Most SEMs in current use still use photographic recording as the main (or only) means of image recording. With good-quality, flat-bed scanners or digital cameras, it is possible to convert these photographs into digital form for analysis. In a digital image, information is recorded as a series of small pixels as an intensity (commonly in the range 0–255). Pixels are electronically digitized arrays or squares defined by two spatial coordinates: (X,Y). The human eye can discriminate between only about 16 gray levels; therefore, digital images have a gray-level resolution at least an order of magnitude greater. The number of pixels per image can vary, but a common standard is an array of 512×512 pixels, although newer systems use 1024×1024, or more, pixels.

BSE images may also be acquired directly without going through the photographic stage. Most SEMs have facilities for TV rates of scanning. The signal thus acquired, provided that it is a standard CCTV scan, may be interfaced directly to a digital frame store in place of the digitizing camera. Images acquired in this manner are inherently noisy, particularly if they are magnified above 1000 times. Improvements in the signal to noise ratio are possible by the use of filtering, which may be achieved by integrating the image over several passes before final capture. Although many electron microscopes will not have digital image capture facilities in their own right, some will have X-ray mode facilities either as an energy-dispersive or a wavelength-dispersive system. Most systems sold since about 1985 can not only acquire X-ray spectra information but also digitally map the distribution of chemical elements in a specimen. For reliable results, SEMs must be calibrated at periodic intervals in terms of both pixel and gray-level resolution.

There are several advantages – such as greater discrimination of intensity and the ability to process the image readily to enhance particular details – to using digitally acquired images rather than using digitized photographic images (Table 9.1). Also, photography itself introduces a form of analogue image processing (contrast enhancement, etc.) over which the operator has no control in subsequent analysis. Furthermore, although digital cameras may digitize the image in a specified range (e.g., 0–255), many will not be able to discriminate that number of discrete gray levels; in extreme cases, only 100 effective gray levels may be present.

Digitally acquired images do have some disadvantages. For example, the use of digital images requires more careful thought with respect to choice of magnification than does photographic recording. The magnification should be chosen to ensure that each feature is covered by at least three pixels (Tovey et al., 1995), and preferably five. Unlike photographic images, digital images have a limited scope for enlarging the useful magnification. Thus, the need to ac-

TABLE 9.1

Advantages and disadvantages of different methods of digital image capture

Indirect digitization through intermediate photograph	Direct digitization from SEM

Advantages

1. Can be used in all laboratories (not all have direct digitizing facilities attached to SEM).	1. Requires specialized interface equipment, although many new microscopes now have such facilities.
2. Photographs from old specimens that no longer exist can be analyzed.	2. On-line prefiltering of the image is possible – an advantage for noisy images.
3. Photographic film can be enlarged several times before photographic grain becomes a limitation. Potentially the spatial resolution of the image will be higher than for directly acquired images. The initial capture magnification is therefore not critical.	3. No additional geometric distortion is introduced.
	4. Only limited photography is needed (for final publication).
	5. Improved gray-level discrimination is possible.
	6. The amount of photographic work needed is greatly reduced.

Disadvantages

1. The print contrast required for direct visual interpretation is usually not the best for digital purposes and dual printing may be necessary.	1. The resolution of the digitized image is very dependent on the initial magnification, as no *useful* enlargement is possible. The operating magnification of the SEM must be chosen carefully at the outset.
2. Gray-level resolution of photographs will rarely be as good as direct digitization, and even when the range is 0–255, gray levels as low as 100 are usually used.	2. Digitally stored images are rarely of as good quality as a direct photograph for final reproduction (although this situation is improving rapidly).
3. Photographs must be processed before digitizing and therefore more photographs than necessary may be captured to ensure adequate gray-level contrast.	
4. Requires at least one intermediate photographic stage, and conditions are less easy to standardize in the development.	
5. Camera lens and enlarge lens distortion add to the geometric displacement of images	

TABLE 9.1 – (continued)

Advantages and disadvantages of different methods of digital image capture

Indirect digitization through intermediate photograph	Direct digitization from SEM

Disadvantages (*continued*)

(although this can be minimized by using the same lens for both photography and printing).

6. Photographic enlargement may introduce further errors of scale.

7. The photographic process introduces further noise.

8. If a camera is used then illumination of the photograph may not be uniform (this is less of a problem with flat bed scanners, but the gray-level resolution of many of these is poor).

9. It is difficult to ensure precisely the same orientation of the photograph during digitization as in the microscope. This may be important for some applications.

quire digital images at a range of magnifications is even more important than doing the same thing for photographic recording.

In spite of the many advantages of direct digital acquisition, digital image acquisition from photographs will likely remain a key method for many workers for at least the next 10 years. When acquiring digital images from photographs, care must be taken to ensure uniform illumination of the print, and, if possible, compensation for nonuniform illumination should be incorporated into all analyses. Photographic prints that have been produced for qualitative interpretation or publication are likely to be too contrasty for image analysis; prints should instead be produced on paper of the softest grade.

Digital images may be stored in a variety of formats, the choice of which will depend upon the application, the precision of the analysis, and data storage requirements. In some cases, it is necessary to convert from one format to another (e.g., image capture may take place on a different system from image analysis). Four formats may be available: byte, integer, real, complex. Some systems do not allow the full range of formats. Byte formats are compact, but care must be exercised during image processing to ensure that the maximum (255) and minimum (0) are not exceeded. Storage of data may be in binary or ASCII form. For processing images where large areas have the same intensity, data compression is possible using run-length encoding. Although many commercial software packages are able to convert images from one format to another, many of these give approximate conversions only. Digital images can require significant amounts of storage space. Storage is less of a problem than it used to be; however, any laboratory equipped for image analysis must have facilities to archive images on optical disks, CDs, Exobyte tapes, or some other equivalent medium.

PROCESSING FOR IMAGE ENHANCEMENT

Before image analysis begins, it is commonly necessary to process the digital image to improve its quality through edge enhancement or other techniques. There are three basic types of processing operations: (1) The intensity of each pixel in the original image is modified in a way that relies solely on the original value at that pixel. The resulting image is unaffected by the intensity at neighboring pixels. A typical example would be contrast stretching (maximizing the difference between black and white areas) or a logarithmic operation. (2) The intensity of the pixel in the output image involves not only the pixel in question but also the surrounding ones. Various combinations of pixels may be used, but it is common, although not essential, to have symmetric arrays (e.g., the nine pixels in a 3 × 3 array). The relative weighting by which each of the pixels contributes to the output image may vary (e.g. equal weighting or differential weighting). The terms "kernel" and "structuring element" are used to describe the exact nature of the processing array. (3) The intensity of the pixel in the output image depends upon the intensities of the corresponding pixel in two or more input images. The operations are normally done on a pixel-by-pixel basis.

A variety of processing operations can be used to enhance digital images. These include (1) contrast stretching, (2) saturation (either or both of the black and white areas become saturated at the minimum and maximum intensities, respectively), (3) truncation (similar to saturation except that only one end of the range is saturated), (4) reversal of the contrast of a photographic image (e.g., solid matter that normally appears bright in BSE images becomes dark, whereas the pores become white), (5) conversion of intensities by replacing each pixel value by the logarithm of the original intensity, (6) rotation of the image into a different orientation, (7) cutting out a subregion of an image, (8) binary segmentation or thresholding (conversion of the raw gray-level image into a binary image where features of one class are white and are encoded unity and the remaining features are black and encoded zero; the aim of segmentation is to detect the optimum threshold that will define the greatest number of edges in the image), (9) reclassification (or mapping) the intensities in the original image to another value (a more general extension of item 8, (10) arithmetic manipulation of images (exponentiation, trigonometric functions, addition, subtraction, multiplication, division), (11) image sharpening (by adding information from the detection of edges to the original image) and blurring (particularly in the computation of the local background intensity), (12) image restoration, which attempts to recover the form of an image before degradation (e.g., correcting for the effects of nonuniform illumination of photographic images by capturing a blank, featureless image and subtracting this image from all images captured during the session), and (13) Fourier methods (use of Fourier transforms of an image may provide more information about the spacing and orientation of features in the original image than is possible from direct observation). Details of these enhancement techniques, many of which are quite advanced and complex, may be found in Gonzalez and Wintz, 1987; Guan and Ward, 1989; Hounslow and Tovey, 1992; Kohler, 1981; Razaz et al., 1993; Tovey and Hounslow, 1995; and Tovey et al., 1994b.

Some commercial feature analysis computer packages may not have all the desirable features needed for a comprehensive image analysis, but it is nevertheless possible by use of additional image processing to overcome many of the deficiencies in these programs. Examples of additional preprocessing that may be required are: (1) filling holes within particles, (2) removing small isolated particles, (3) selective line thickening, and (4) separating features in binary images. Some image-processing packages include facilities for such operations; for others, it may be necessary to write special routines if these facilities are required frequently. It is always possible to use manual editing to fill holes within particles, but this can become tedious if many particles are so affected. Using standard feature analysis packages, it is normally possible to exclude small isolated particles or, alternatively, a hit-and-miss transform may be used. For this purpose, a simple kernel (the exact nature of the processing array) is passed over the image. Where there is an exact match, the corresponding pixel in the output image is coded unity; otherwise it is coded as zero. Subtracting this output image from the original then produces one in which isolated pixels have been removed. Using a similar procedure, but with multiple passes, it is possible to remove the ends of lines, pairs of pixels, etc. More advanced applications allow selective line thickening, which is often needed before feature analysis methods can be used.

IMAGE ANALYSIS

Introduction

To accomplish the tasks discussed above, digital images obtained by direct acquisition methods or by digitizing BSE photographs must be analyzed after enhancement by using image analysis computer packages. For example, mineralogic analysis is performed by separating gray-level images into separate binary images for each gray level and identifying the mineral represented by each gray level (Ball and McCartney, 1981; Pye, 1984a; Dilks and Graham, 1985). The computer is much better than the human operator in discriminating between shades of gray. However, although pattern recognition is relatively simple for the human operator, it is difficult to develop robust algorithms for textural recognition by the computer. There are opportunities for combining the best attributes of both in semiquantitative methods. A particular example is the extension of the hand-mapping method of McConnochie (1974) to allow both feature delineation by the observer and computation of the areas by the computer (Tovey et al., 1989). For most image analysis methods, it is necessary to preprocess the images as described in the preceding section.

Standard Feature Analysis Packages

With a binary image it is a simple matter, for example, to estimate porosity by measuring the proportion of pixels that are coded as unity and those coded zero (e.g., Pye, 1984a; Tovey and Hounslow, 1995). Most image-processing and image-analysis computer packages also have facilities to measure the size and shape of particles or other features of the images (Bisdom and Schoonderbeek, 1983; Petruk, 1989a; Smith, 1989). These analyses use binary images, but the requirements for segmentation by gray level are more stringent and more difficult to achieve than those for porosity analysis because no feature must touch any of its neighbors. During analysis, each individual feature is labeled by a unique code (i.e., a labeled image is generated in which all the pixels of each separate feature are coded to a specific value). Most good packages allow some editing of features (e.g., exclude features less than a particular size, exclude those touching an edge, exclude those with holes, exclude those whose shape is in a given range). Most feature analysis packages generate information on many key parameters on each particle, such as size, shape, and orientation (Table 9.2).

Problems of Connectivity

In all standard image analysis packages, implicit decisions are necessary with regard to connectivity between foreground and background regions. **Four-point connectivity** is used when foreground regions are considered as continuous, provided that pixels touch horizontally or vertically. **Eight-point connectivity** includes also the situation in which foreground pixels can touch diagonally. Therefore, it is important to recognize which criterion is in use in a particular package. Four-point connectivity in the foreground automatically implies eight-point connectivity in the background, and vice versa.

Feature Chord Length Measurement

Using a binary image, it is possible to estimate the relative thickness of features present. This can be done by measuring the mean intercept of solid features in the horizontal directions in the original image. At the same time, the mean thickness of pores is determined. An alternative is to measure the vertical chord length, inasmuch as packing variations may cause the effective thicknesses in the two directions to vary. Algorithms to estimate the mean chord lengths are simple. Some features will inevitably touch the edge of the image, and decisions must be made regarding these. They may be either neglected (as their true thickness is unknown); included, in which case they will bias the results to smaller

TABLE 9.2

Typical parameters that are estimated in many feature analysis packages (more advanced parameters may be derived from these, including an estimate of how "lumpy" particles are)

Parameter	Comments
Particle identifier	A unique number for each particle.
X and Y coordinates of a reference point	Reference point often extreme bottom left-hand corner of rectangle enclosing particle.
Number of holes	Number of "holes" within the particle.
Contact flag	Flag is set if particle touches an edge of the image, otherwise unset.
Minimum X Maximum X Minimum Y Maximum Y	These particles define the enclosing rectangle, and are used if an individual particle is to be "extracted" from the image.
Horizontal Feret diameter	
Vertical Feret diameter	
2 inclined Feret diameters	At 45°.
Horizontal projection Vertical projection	These will be the same as the corresponding Feret diameters for simple shapes.
Perimeter	Measured in pixels.
Area	Measured in pixels.
X and Y coordinates of center of area	For some shapes this may lie outside particle.
Maximum and minimum principal moments of area	Useful to see how mass of particle is distributed.
Orientation	Inclination of the axis of the longest direction to the x axis.
Circularity	A simple measure of shape being a function of the area and perimeter. Problems can arise when circularity of a particle consisting of a single pixel is measured.

mean chord lengths; or included only if they satisfy a given criterion (e.g., they are longer than the mean chord length determined without them). Mean chord length determinations are a particularly efficient way to estimate mean feature sizes in a binary image.

Orientation Analysis:
Intensity-Gradient Analysis Methods

The image-processing and analysis techniques so far discussed in this chapter have covered the initial processing of images and the problems associated with segmentation by gray level. Some techniques require little or no preprocessing and thus are ideal for fully automated analysis. A key area here is orientation analysis, in which the alignment of features is studied. As a result of sediment loading or transport of particles by water, platy or elongated particles may become aligned in one direction or another. Similarly, there may be patterns on a feature that are aligned in one direction that are different from those on other features. Thus, it becomes possible to segment images on the basis of their orientation characteristics.

The intensity gradient methods make use of the initial gray-level image and avoid the typical problems of image segmentation. These methods are ideally suited to the study of alignments in fine-grained materials, but they are less suitable for large or nearly equidimensional features. In many sediments consisting of a wide range of particles sizes, it is necessary during orientation analysis to combine the advantages of the intensity gradient analysis with those of other techniques such as the multispectral methods discussed in the next section (Tovey and Krinsley, 1992).

Features and particles in images are bounded by their edges where the intensity is changing abruptly from bright to dark or vice versa. The greatest rate of change in intensity will be orthogonal to the edge of the feature. Intensity-gradient analysis methods attempt to measure the direction of this change and hence define the orientation at each pixel in an image. This approach was suggested by Unitt (1975); it was first used in the study of soils and sediments by Tovey (1980) and subsequently by Tovey and Sokolov (1981) and Smart and Tovey (1988). In its simplest form, we may consider only three pixels, a central pixel (numbered 0 in Fig. 9.1) and two adjacent pixels (labeled 1 and 2). If the intensities at the three pixels 0, 1, and 2 are I_0, I_1, and I_2, respectively, then the gradient of the change in intensity in the X direction ($\Delta I/\Delta x$) is $(I_1 - I_0)/h$, where h is the distance between pixels and the corresponding gradient in the Y direction ($\Delta I/\Delta y$) is $(I_2 - I_0)/h$. The direction of the maximum gradient (θ) is then given by

$$\theta = \tan^{-1}\left(\frac{\Delta I/\Delta y}{\Delta I/\Delta x}\right) \tag{9.1}$$

For some purposes it is helpful to determine the magnitude (M) of the intensity gradient vector. This may be used for detecting edges in an image. M will be relatively high at an edge and low or zero in regions of no contrast. This is defined as:

$$M = \sqrt{(\Delta I/\Delta x)^2 + (\Delta I/\Delta y)^2} \tag{9.2}$$

For additional information on the techniques of orientation analysis, see Smart and Tovey, 1981–1982; Tovey and Smart, 1986; Zuniga and Haralick, 1987; Smart and Tovey, 1988; Tovey et al., 1989, 1990; Smart et al., 1990; Sokolov, 1990;

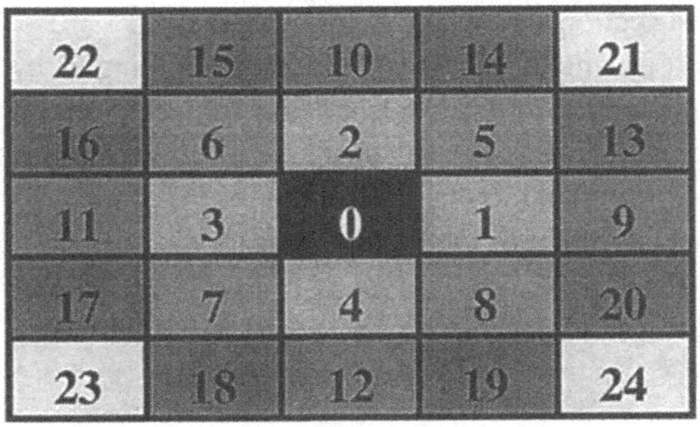

Figure 9.1. Pixel numbering system used in intensity-gradient orientation analysis (see Smart and Tovey, 1988). The sequence is chosen for efficient computation and to provide groupings of symmetric arrays of pixels. The shadings represent pixels at different distances from the center.

22	15	10	14	21
16	6	2	5	13
11	3	0	1	9
17	7	4	8	20
23	18	12	19	24

Tovey and Martinez, 1991; Smart and Leng, 1993; and Tovey et al., 1995. Mardia (1992) is a useful reference for statistical analysis of data obtained.

Figure 9.2 shows two typical examples of BSE images of resin-impregnated samples of consolidated clay. One sample (A) has a predominant orientation direction running from top left to bottom right; in the other (B), there are local regions of orientation, but it is unclear if there is an overall general orientation within the image as a whole. Figure 9.2C is a domain-segmented image corresponding to Figure 9.2A, and Figure 9.2D is the domain-segmented image corresponding to Figure 9.2B. The orientations determined from these figures can be conveniently displayed as rosette diagrams (Figs. 9.2E and 9.2F). In most BSE images of fine-grained sediments, the rosette diagrams are elliptical in shape. It is permissible to determine the best-fitting ellipse, from which the direction of preferred orientation may be determined and, from the lengths of the major and minor axes, describe an index of anisotropy (I_a) defined as

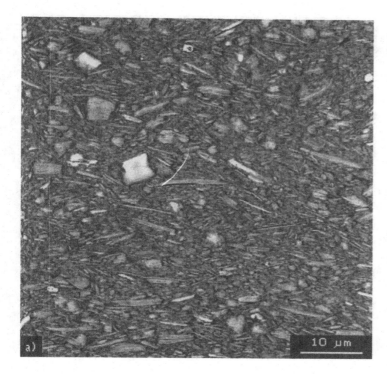

a)   10 µm

Figure 9.2 (a). Example of orientation analysis. BSE image showing two distinct, contrasting microfabrics.

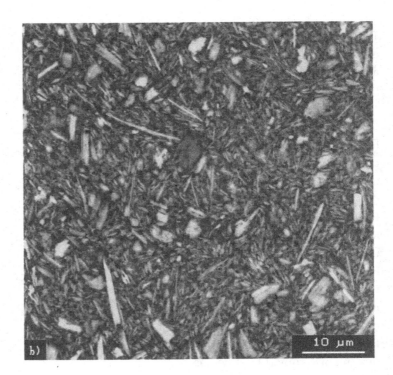

Figure 9.2 (b). Example of orientation analysis. BSE image showing two distinct, contrasting microfabrics.

Figure 9.2 (c). Domain-segmented image of (a); this is the final stage of orientation analysis and automatically defines regions with a particular general orientation. In this example, only four orientation classes are used: vertical, horizontal, and two inclined at 45°. A fifth class for randomly oriented microfabrics is used for regions where there is no dominant orientation (see Tovey et al., 1995, for full description of the method).

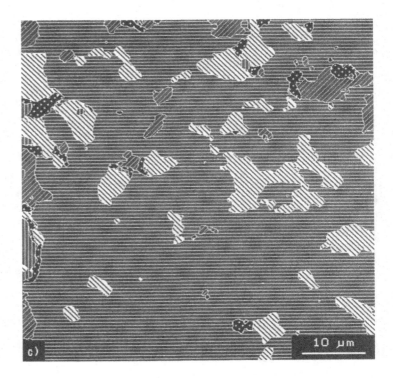

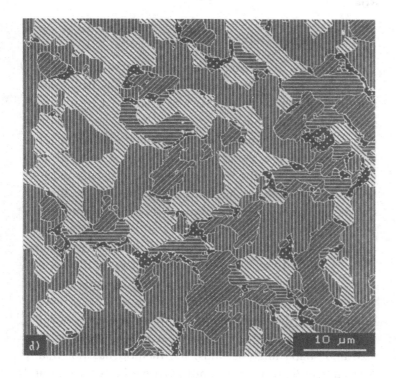

Figure 9.2 (d). Corresponding domain-segmented image for (b).

$$I_a = 1 - \frac{\text{length of minor principal axis}}{\text{length of major principal axis}} \qquad (9.3)$$

This index ranges in value from zero for a random structure to unity for a perfectly aligned sample, with most samples falling in the range 0.05–0.85.

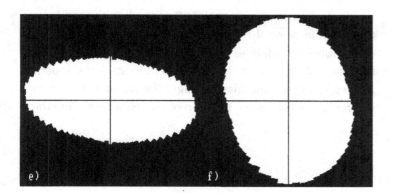

Figure 9.2 (e) and (f). Rosette diagrams for (a) and (b), respectively.

161

Multispectral Methods

Introduction and General Description of Method.

The BSE image is just one of many that may be acquired from a particular area in the SEM. SE images, CL images, and specimen-current images, as well as X-ray maps for each chemical element, may also be obtained. With the large number of different image types available, it is possible to use multispectral methods in image analysis of soils and sediments. Some work has already been done using multispectral methods on optical sections of soils (e.g., Protz et al., 1992, and Terribile and Fitzpatrick, 1995), and also a limited amount of work in SEM studies of sediments (Tovey and Krinsley, 1991, 1992; Tovey et al., 1992a, 1992b, 1994a).

Multispectral methods may be used to separate different mineral species within an image in a way that cannot be done reliably by other methods. Once this has been done, each mineral species can be examined separately for particle size and shape analysis; thus, the method has potentially more power than do more traditional methods for size and shape analysis. The methods rely on the presence of characteristic differences in the intensity ranges of the various minerals in one or more of the different images. In theory, with ideal images and in the absence of noise, it might be possible to separate mineral species using the BSE images alone (Ball and McCartney, 1981; Pye, 1984a; Dilks and Graham, 1985); however, separation will work for only a limited combination of minerals. For example, Tovey and Krinsley (1991) demonstrated that it was not possible to separate the common minerals quartz and K-feldspar by this means.

Digital SE and BSE images can be acquired with sufficient freedom of noise in a matter of 10 seconds, although longer times can improve the image. X-ray images require a significantly longer dwell time on each pixel; therefore, the recording time will be longer. It is convenient to stack the separate images as a single multilayer image. The *X* and *Y* dimensions still represent the spatial location of points, whereas the third dimension represents different spectral bands (e.g., layer 1 might be the BSE image, layer 2 might be the X-ray map of Mg concentration, layer 3, the aluminum concentration, and so on). Using multilayer images means that a single disk file is used to store all the information rather than the information being dispersed through separate files for each image.

Selection of Training Areas and Classification

For most multispectral processing it will be necessary to identify small training areas that are representative of the minerals present in the image. The computer will then use the statistical distributions of gray levels within the training areas to classify the whole image. Various criteria may be set, such as a forced classification, in which the whole image is classified according to one of the identified minerals, or, alternatively, a level of uncertainty may be included that will leave some areas unclassified, commonly at the boundaries between features where there is no clear information about the true classification.

As an example, Figure 9.3A shows a sample of the Whitby Shale, Yorkshire, U.K., in which quartz, feldspar, and pyrite grains are embedded in a matrix of finer clay minerals. Smaller particles of other minerals are also present. The corresponding X-ray (chemical-element) maps are shown in Figure 9.3B. Quartz is identified in regions with high silicon concentration and an absence of other elements. Feldspar grains are characterized by moderate concentrations of silicon and aluminum, together with potassium; high concentrations of iron and sulphur indicate pyrite. Such knowledge of chemical composition of the minerals is required in the selection of training areas. Typically, training areas should have a minimum size of about 100 pixels to ensure adequate statistical data; the actual areas used in this example are indicated in Figure 9.3A.

Regions may be present within many images that represent the background clay matrix; one training area should be selected as being representative of such an area. In some

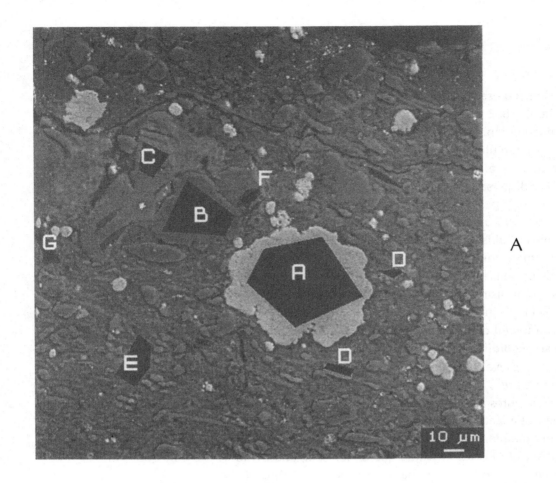

A

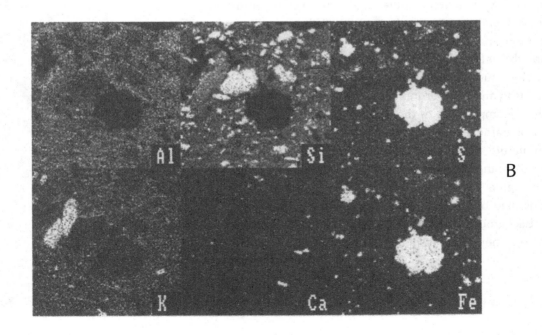

B

cases, two or even three matrix regions may be selected to discriminate between matrix consisting primarily of clay and matrix that may include amorphous iron.

Once the training areas have been selected, the statistics of intensity variations within the respective training areas within each layer are computed and stored in the covariance matrix. At the time of this computation, it is important to select the relevant layers for the classification. Thus, if two layers give essentially the same spatial information, there is little point in including both in either the statistical analysis or the subsequent classification. Similarly, for some X-ray maps the distribution of a particular element is uniform and will provide no additional information in any discrimination process; such layers can be safely omitted. Careful selection and discarding of image layers reduce the time needed for image classification and also minimizes ambiguities in classification. Areas in which the intensity of one particular layer is constant throughout should be avoided, as zeros will be present in the covariance matrix, leading to singularity problems. Tovey and Krinsley (1991) indicate a method to overcome this difficulty where such problems are unavoidable.

Image Classification

Once the covariance matrix from the training areas has been computed, the data may be used for classification. Several different classification methods are available, but the **maximum likelihood classifier** appears to be the most robust for use in image analysis in the earth sciences. The classification process uses information from the covariance matrix to decide on which class to allocate every pixel in the output image. Different levels of precision may be set, ranging from a forced classification (in which all pixels are forced to the nearest possible class) to methods in which classification is done only when the statistics indicate that a particular pixel falls within a predetermined range from the central point of the class in question. Thus, a 95% level may be set that would indicate that a particular class would be allocated

Figure 9.3 (opposite). (A) Backscattered electron (BSE) image of Whitby Shale showing training areas used in multispectral analysis to generate images in Figure 9.4. Training areas: A = pyrite; B = quartz; C = K-feldspar; D = calcite (two areas); E = clay matrix; F = inclusions in quartz grain; G = void. (B) X-ray maps of selected elements for the BSE image shown in (A). The lighter areas represent areas with a high concentration of the relevant element. Al = aluminum; Ca = calcium; Fe = iron; K = potassium; S = sulphur; Si = silicon.

provided that the statistics at a particular pixel fell within two standard deviations of the mean of the class in question. For many purposes, it is desirable to leave a degree of uncertainty; experience shows that threshold levels of 90%, 95%, and 98% seem suitable levels to adopt. See Tovey and Krinsley (1991) for details.

As classification proceeds, the pixels in the output image are coded as 1 if the statistics suggest that the particular pixel is most like the training area for mineral 1. Similarly, pixels representing mineral 2 will be coded as 2, and so on. For classification (other than forced classification), there will remain some pixels that are coded as 0. These pixels represent the areas that fall at the extremes of classes (i.e., at greater distance from the class mean than the specified threshold). In most cases, these regions of uncertainty appear at the boundaries of particles or, in some cases, where no relevant training area has been selected. Figure 9.4 shows examples of classifying the image in Figure 9.3 using both a forced classification and a threshold level such that each classified pixel is within 95% of the mean of a particular class.

An estimate of the proportions of each mineral present may be obtained by computing the relevant areas of each class. The histogram in Figure 9.5 shows the results obtained by both forced classification and classification at the 95% level. Provided that reasonable care is taken with the selection of training areas, the normalized relative proportions of minerals present are very similar irrespective of the threshold confidence level set, and the variation is comparable to other methods of mineralogical analysis (e.g., X-ray diffraction).

Combination Methods and Other Techniques

Many useful applications may be achieved using image processing and analysis as described in this chapter. There are, however, many other methods that involve combinations of analysis and/or processing that do not fit easily into any of the categories described in the previous sections.

Figure 9.4 (opposite). (a) Classification of image in Figure 9.3 using training areas specified. Classification was forced to one of the specified classes at all pixels in the image. A limited amount of image processing was done to remove very small (<1 mm) holes or particles. (b) Classification of image in Figure 9.3A using training areas specified. A 95% level was set and regions of uncertainty are left unclassified. Most unclassified areas are at boundaries between features or small holes in features. Some are larger regions (e.g., at A), suggesting that at least one additional mineral is present, which could be defined by an additional training area at A.

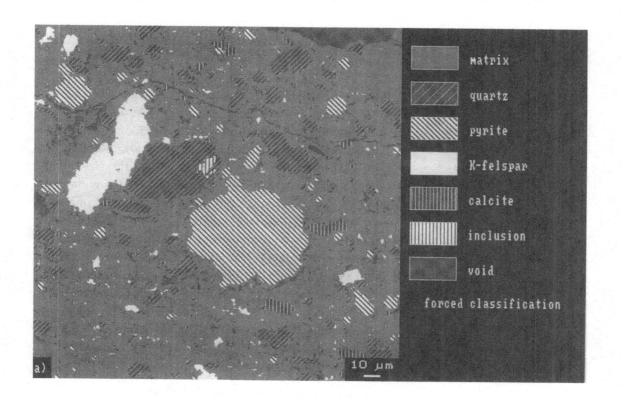

a)

	matrix
	quartz
	pyrite
	K-felspar
	calcite
	inclusion
	void

forced classification

10 µm

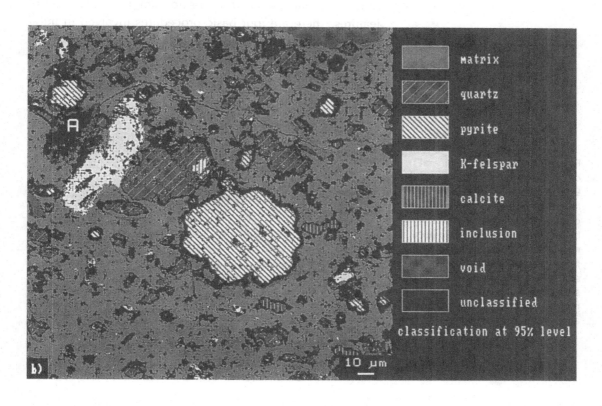

b)

	matrix
	quartz
	pyrite
	K-felspar
	calcite
	inclusion
	void
	unclassified

classification at 95% level

10 µm

Many of these methods involve several stages. For example, it is possible to study how porosity within a sample varies from domain to domain (e.g., Tovey, 1995, and Tovey et al., 1994b, 1995). Also, multispectral methods may be combined with domain segmentation to study the size, shape, and orientation of particles in rocks that have a wide range of grain sizes. These methods allow processing of images in two stages. The first stage examines the larger particles; the second examines the fine-grained matrix. Such images are referred to as mineral-segmented images (Tovey and Krinsley, 1992; Tovey et al., 1992a). The use of Fourier transforms and image analysis to study and classify microfabrics in soils, sediments, and desert varnish is also a potentially useful technique (Tovey, 1971; Tovey and Wong, 1974, 1978; Tovey and Sokolov, 1981; Sokolov, 1990; Derbyshire et al., 1992; Tovey and Sokolov, in press).

Mathematical Morphological Methods

Morphological methods can provide direct, overall information about feature- or void-size distribution and particle roughness with the need to ensure complete separation of features. There are four basic procedures involved in these morphological operations: (1) erosion (a single layer of pixels is stripped from the surfaces of foreground features); (2) dilation (a single layer of pixels is added to the surface of

Figure 9.5. Histogram of mineral distributions in Figures 9.4A and 9.4B generated by using both forced classification and classification at the 95% level. In the latter case, the data have been normalized as percentages of features actually classified. The two sets of data are consistent, suggesting that analysis of micromineralogical composition to about ±1% is possible.

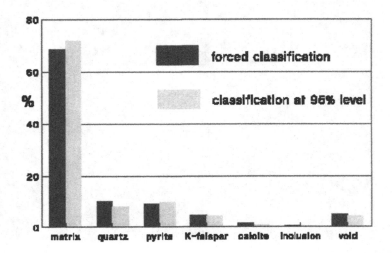

foreground features); (3) opening (consisting of a single erosion followed by a single dilation); (4) closing (consisting of a single dilation followed by a single erosion). In an opening, a binary-source image is first combined with a structuring element to produce an image whose features are reduced by a factor related to the size of the structuring element (erosion). The resulting image is then dilated with an identical structuring element to increase the size of the features (dilation). During this process, features smaller than the size of the eroding structuring element are lost from the image, as there is no seed point for the subsequent dilation. As a result, measurements before and after opening provide the area of the features lost. Opening algorithms are useful for (solid) feature-size analysis, whereas the reverse process (closing) is useful for pore-size distributions. Alternatively, a closing operation can be converted to an opening sequence by reversing the contrast of the original image. For example, Ehrlich et al. (1984) describe the application of a sequence of openings to study the pores in reservoir complexes in sandstones; it is a simple matter to extend the technique to study porosity and particle size distributions in BSE images of sediments (e.g., Tovey, 1995). Extensions of the basic morphological methods making direct use of gray-level images are now possible (see also Prod'homme et al., 1992, and Tovey and Dent, in press).

INTERPRETATION AND APPLICATION

Image analysis can be applied to a number of tasks that involve measurement of features in sediments and soils such as particle size, shape, and orientation. Ultimately, however, these feature measurements must be interpreted and applied to the solution of real problems. In the earth sciences, the most significant applications in the past have been in the fields of mineralogy and petrology.

Considerable work has been carried out to characterize

the mineralogy of sandstones and shales (e.g., Dilks and Graham, 1984; Krinsley et al., 1983; Minnis, 1984; Pye, 1984a; Pye and Krinsley, 1984; White et al., 1984), including automated analysis of relative mineral abundance. For example, Dilks and Graham (1984) used an electron probe microanalyzer interfaced to an image analyzer to study sandstone petrography, including a comparison of mineralogy modes determined by electron-probe point counting and by image analysis. Some minerals can be identified by their gray levels, provided that the composition of the sample is not complex. Those that cannot be so discriminated must be identified on the basis of elemental composition by the multispectral methods described above. Binary images produced from segmentation of minerals with similar gray levels are used as a mask for X-ray analysis. This also allows a pixel-mapping program to be run on these minerals, resulting in the classification of the proportions of each mineral.

One of the main problems with the above method of mineral classification by image analysis is that some minerals have similar BSE coefficients and cannot be discriminated reliably on the basis of gray level. Multispectral methods, as described in this chapter, may be necessary to effect discrimination. A second problem arises with shales or other fine-grained rocks in which resolution and discrimination are limited by the size of the excitation volume in minerals of low average atomic number. Difficulties may also arise when analyzing samples that display a large range of grain sizes – from sand size to clay size. It can be helpful to use the mineral maps described above for a mask to "separate" the larger grains in an image from the fine-grained matrix. Two images are thus generated: one that is eminently suited to standard feature analysis (i.e., the larger grains), and one that may be processed by using the orientation analysis method described above.

Considerable work has also been done to study the porosity of sediments and soils by image analysis techniques, including measurement of pore size and total pore space, as well as description and qualification of the shapes of pores (e.g., Cohen and Anderson, 1985; Ehrlich et al., 1984; Pye, 1984a; Ruzyla, 1986; Stuart et al., 1987; Tovey and Hounslow, 1995). Much of this work has focused on characterization of pores in hydrocarbon reservoir rocks; however, some work has also been done on the porosity of soils. The use of image analysis to characterize the porosity and microfabrics of soils, including orientation of minerals in soils, has been investigated extensively by Tovey and coworkers (e.g., Tovey, 1995; Tovey and Dent, in press; Tovey and Hounslow, 1995; Tovey and Krinsley, 1992; Tovey and Sokolov, in press; Tovey et al., 1990; Tovey et al., 1992b; Tovey et al., 1992c; Tovey et al., 1992d; Tovey et al., 1994a; Tovey et al., 1994b).

Other applications of image analysis in the earth sciences include: (1) acquisition and segmentation of images for the purpose of characterizing shapes of particles by Fourier grain-shape analysis, fractal analysis, or other techniques (e.g., Ehrlich et al., 1984; Smith, 1989); (2) measurement of mineral matter in coal (e.g., Gottlieb et al., 1989); (3) characterization of the dispersion of particles (Shehata, 1989); (4) mineral beneficiation (Petruk, 1989b); and (5) integration of geoscientific map data (Bonham-Carter, 1989). For more details of these applications, and of additional applications in the earth sciences, see Petruk (1989c).

CONCLUDING REMARKS

The human eye still commands an advantage over computer systems in the area of pattern recognition, although with the development of new techniques such as fractal dimension and co-occurrence statistical methods, objective pattern recognition and segmentation should become possible within the next 5 to 10 years. However, computer systems are much superior in the areas of gray-level discrimination and objectivity in analysis. Indeed, there are compelling reasons to use a combination of human and computer-based image analysis while pattern recognition undergoes more development.

Subjective interpretation of images can be biased; unless the investigator is experienced, details in parts of an image may be correctly described but it will be difficult to appreciate the effect in the whole image when the structure is complex. It is tempting to concentrate unduly on the brighter features of an image rather than giving comparable weight to the less contrasting parts. Therefore, there is merit in using image analysis to provide an image that is more amenable to interpretation. Thus, image processing and analysis are used not only to obtain quantitative parameters, but also to provide a more rational basis for subjective interpretation.

As image analysis and processing continue to develop and provide ever more exciting research possibilities, there is an important need to ensure that these developments are brought to the attention of earth scientists. Ironically, articles relevant to earth scientists often appear in literature that they rarely consult. One of the major aims of this chapter is to bridge this image analysis information gap between earth scientists and other scientists and to provide insight into the research techniques in image analysis that are currently available.

Bibliography

Agar, S., Prior, D., and Behrmann, J. (1989) Backscattered electron imagery of the tectonic fabrics of some fine-grained sediments: Implications for fabric nomenclature and deformation processes. *Geology*, 17, 901–904.

Allen, D. (1984) A one-stage precision polishing technique for geological specimens. *Mineralogical Magazine*, 48, 298–300.

Allison, P. A., and Pye, K. (1994) Early diagenetic mineralization and fossil preservation in modern carbonate concretions. *Palaios*, 9, 561–575.

Amthor, J. (1993) Combining cathodoluminescence and backscattered electron microscopy in the study of diagenetic carbonates. *J. Geological Education*, 41, 140–143.

Amthor, J. E., and Friedman, G. M. (1992) Early- to late-diagenetic dolomitization of platform carbonates: Lower Ordovician Ellensburger Group, Permian Basin, West Texas. *J. Sedimentary Petrology*, 62, 131–144.

Autrata, R., Jirak, J., Spinka, J., and Hutar, O. (1992) Integrated single crystal detector for simultaneous detection of cathodoluminescence and backscattered electrons in scanning electron microscopy. *Scanning Microscopy*, 6, 69–80.

Bailey, A., and Blackson, J. (1984) Examination of organic-rich sediments structurally maintained using low viscosity resin impregnation. *Scanning Electron Microscopy*, 1475–1481.

Ball, M. D., and McCartney, D. G. (1981) The measurement of atomic number and composition in an SEM using backscattered detectors. *J. Microscopy*, 124, 57–68.

Barker, C. E., and Kopp, O. C., eds. (1991) Luminescence microscopy and spectroscopy: Qualitative and quantitative applications. *SEPM Short Course 25, Soc. Sedimentary Geology*, Tulsa, Okla.

Barrows, M., 1980, Scanning electron microscopy studies of samples from the New Albany Group. *Scanning Electron Microscopy*, 578–585.

Bennett, R. H., Bryant, W. R., and Hulbert, M. H., eds. (1991a) *Microstructure of Fine-Grained Sediments, from Mud to Shale.* Springer-Verlag, New York.

Bennett, R. H., O'Brien, N. R., and Hulbert, M. H. (1991b) Determinants of clay and shale microfabric signatures: Processes and mechanisms. In Bennett, R. H., Bryant, W. R., and Hulbert, M. H., eds., *Microstructure of Fine-Grained Sediments.* Springer-Verlag, New York, pp. 5–32.

Berner, R. A. (1971) *Principles of Chemical Sedimentology.* McGraw-Hill, New York.

Bisdom, E. B. A., and Schoonderbeek, D. (1983) The characterisation of the shape of mineral grains in thin sections by Quantimet and BSEI. *Geoderma*, 30, 303–322.

Bisdom, E. B. A., and Thiel, F. (1981) Backscattered electron scanning images of porosities in thin sections of soils, weathered rocks and oil-gas reservoir rocks using SEM-EDXRA. In Bisdom, E. B. A., ed., *Submicroscopy of soils and weathered rocks.* Centre for Agricultural Publishing and Documentation, Wageningen, pp. 191–206.

Bishop, H. E. (1966) Some electron backscattering measurements for solid targets. In Castaing, R., Deschamps, P., and Philibert, J., eds., *Optique des Rayons X et Microanalyse.* Herman, Paris, pp. 153–158.

Blatt, H. (1982) *Sedimentary Petrology.* W.H. Freeman, San Francisco.

Boggs, S., Jr. (1992) *Petrology of Sedimentary Rocks.* Macmillan, New York.

Boggs, S., Jr., and Seyedolali, A. (1992) Diagenetic albitization, zeolitization, and replacement in Miocene sandstones, Sites 796, 797, and 799, Japan Sea. In Pisciotto, K., Ingle, J. C. Jr., Von Breymann, M. T., and Barron, J., et al., eds., *Proceedings of Ocean Drilling Program, Scientific Results*, 127/128, Pt. 1, College Station, Tex., 131–151.

Boggs, S., Jr., and Seyedolali, A. (1993) Provenance of Miocene sandstones from ODP sites in the Japan Sea. *J. Sedimentological Soc. Japan*, 38, 5–24.

Bonham-Carter, G. F. (1989) Integration of geoscientific maps using image analysis and geographic information systems. In Petruk,

W., ed., *Image Analysis in Earth Sciences. Mineralogical Association of Canada, Short Course Handbook*, 16, 155–156.

Bornhold, B. D., and Giresse, P. (1985) Glauconitic sediments on the continental shelf off Vancouver Island, British Columbia, Canada. *J. Sedimentary Petrology*, 55, 653–664.

Brady, S., and Krinsley, D. (1990) Diagenetic changes associated with foraminiferal remains in Texas Gulf Coast sediments. *Carbonates Evap.*, 5, 1–12.

Budd, D. A., and Hiatt, E. C. (1993) Mineralogical stabilization of high-magnesium calcite: Geochemical evidence for intracrystal recrystallization within Holocene porcellaneous foraminifera. *J. Sedimentary Petrology*, 63, 261–274.

Burton, J., Krinsley, D., and Pye, K. (1987) A back-scattered electron imaging study of authigenic chlorite and kaolinite in Texas Gulf Coast sediments. *Clays and Clay Minerals*, 35, 291–296.

Carlos, B. A, Chipera, S. J., Bish, D. L., and Craven, S. J. (1993) Fracture-lining manganese oxide minerals in silicic tuff, Yucca Mountain, Nevada, USA. *Chemical Geology*, 107, 47–69.

Causton, B. (1988) The hazards associated with embedding resins. *Microscopy and Analysis*, January, 19–21.

Chalcraft, D., and Pye, K. (1984) Humid tropical weathering of quartzite in southeastern Venezuela. *Z. Geomorph. N.F.*, 28, 321–332.

Chaudhuri, A., Chanda, S., and Dasgupta, S. (1994) Proterozoic glauconite peloids from South India: Their origin and significance. *J. Sedimentary Research*, A64, 765–770.

Choquette, P. W., and James, N. P. (1987) Diagenesis in limestones-3. The deep burial environment. *Geoscience Canada*, 14, 3–35.

Cohen, M. H., and Anderson, M. P., (1985) The chemistry and physics of porous media. *The Electrochemical Society*, Pennington, New Jersey.

Cook, S., and Parker, A. (1988) Compositional analysis of sedimentary rocks from SEM/X-ray imaging of polished thin sections. *Microscopy and Analysis*, July, 24–25.

Darlington, E. F., and Cosslett, V. E. (1972) Backscattering of 0.5–10 keV electrons from solid targets. *J. Phys.*, D 5, 1969–1981.

Derbyshire, E., Unwin, D. J., Fang, X. M., and Langford, M. (1992) The Fourier frequency-domain representation of sediment fabric anisotropy. *Computers and Geosciences*, 18, 63–73.

Dijkshoorn, L., and Fens, T. (1992) Sample preparation of sandstone samples prior to the automatic mineral identification analysis. *Scanning and Optical Microscopy*, 20–22.

Dilks, A., and Graham, S. C. (1984) Quantitative compositional and morphological characterization of sandstones by backscattered electron image analysis. In Romig, A. D., Jr., and Goldstein, J. I., eds., *Microbeam Analysis.* San Francisco Press, San Francisco, pp. 149–153.

Dilks, A., and Graham, S. C. (1985) Quantitative mineralogical characterization of sandstones by back-scattered electron image analysis. *J. Sedimentary Petrology,* 55, 347–355.

Dorn, R. I. (1984) Cause and implications of rock varnish microchemical laminations. *Nature,* 310, 767–770.

Dorn, R. I. (1986) Rock varnish as an indicator of aeolian environmental change. In W. G. Nickling, ed., *Aeolian Geomorphology.* Allen & Unwin, London, pp. 291–307.

Dorn, R. I. (1994) Surface exposure dating with rock varnish. In C. Becker, ed., *Dating in exposed and surface contexts.* Univ. New Mexico Press, Albuquerque, N.M., pp. 77–113.

Dorn, R. I., Clarkson, P., Nobbs, M., Loendorf, L., and Whitley, D. (1992) New approach to the radiocarbon dating of rock varnish, with examples from drylands. *Ann. Assoc. Am. Geogr.,* 82, 136–151.

Dorn, R. I., and Dragovich, D. (1990) Interpretation of rock varnish in Australia: Case studies from the arid zone. *Australian Geog.,* 21, 19–32.

Dorn, R. I., Nobbs, M. and Cahill, T. (1988) Cation-ratio dating of rock engravings from the Olary Province of arid South Australia. *Antiquity,* 62, 681–689.

Dorn, R. I., and Oberlander, T. (1981) Rock varnish origin, characteristics and usage. *Zeitschrift fur Geomorph.,* 25, 420–436.

Dorn, R. I., and Oberlander, T. (1982) Rock varnish. *Prog. Phys. Geogr.,* 6, 317–367.

Dorn, R. I., and Whitley, D. (1984) Chronometric and relative age of petroglyphs in the western United States. *Ann Assoc. Am Geogr.,* 74, 308–322.

Douglas, G. R. (1987) Manganese-rich rock coatings from Iceland. *Earth Science Processes and Landforms,* 12, 301–310.

Drake, N. A., Heydeman, M. T., and White, R. H. (1993) Distribution and formation of rock varnish in southern Tunisia. *Earth Surface Processes and Landforms,* 12, 301–310.

Duncumb, P., and Shields, P. K. (1963) The present state of quantitative X-ray microanalysis. Part 1. Physical basis. *Brit. J. App. Phys.,* 14, 617–625.

Dypvik, H., Nagy, J., and Krinsley, D. H. (1992) Origin of the Myklegardfjellet bed, a basal Cretaceous marker on Spitsbergen. *Polar Research*, 11, 21–31.

Ehrlich, R., Brown, P. J., Yarus, J. M., and Przygocki, R. S. (1980) The origin of shape frequency distributions and the relationship between size and shape. *J. Sedimentary Petrology*, 50, 475–483.

Ehrlich, R., Crabtree, S. J., Kennedy, S. K., and Cannon, R. L. (1984) Petrographic image analysis, I. Analysis of reservoir pore complexes. *J. of Sedimentary Petrology*, 54, 1365–1378.

Ehrlich, R., Etris, E. L., Brumfield, D., Yuan, L. P., and Crabtree, S. J. (1991) Petrography and reservoir physics, III. Physical models for permeability and formation factor. *Amer. Assn. Petroleum Geol. Bull.*, 75, 1579–1592.

Ehrlich, R., and Weinberg, B. (1970) An exact method for characterization of grain shape. *J. Sedimentary Petrology* 40, 205–212.

Evans, J., Hogg, A. J. C., Hopkins, M. S., and Howarth, R. J. (1994) Quantification of quartz cements using combined SEM, CL and image analysis. *J. Sedim. Res.*, A 64, 334–338.

Everhart, T. E., and Thornley, R. F. M. (1960) Wide-band detector for microampere low-energy electron currents. *J. Sci. Instrum.*, 37, 246–248.

Folk, R. L. (1960) Petrography and origin of the Tuscarora, Rose Hill and Keefer Formations, Lower and Middle Silurian of eastern West Virginia. *J. Sedimentary Petrology*, 30, 1–59.

Folk, R. L. (1962) Petrography and origin of the Silurian Rochester and McKenzie shales, Morgan County, West Virginia. *J. Sedimentary Petrology*, 32, 539–578.

Fynn, G. W., and Powell, W. J. A. (1979) *The Cutting and Polishing of Electro-Optical Materials.* Adam Hilger, London.

Gedke, D. A., Ayres, J. B., and Denee, P. B. (1978) A solid state backscatter electron detector capable of operating at TV scan rates. *Scanning Electron Microscopy*, 581–592.

Glennie, K. W. (1970) *Desert Sedimentary Environments.* Elsevier, Amsterdam.

Goldstein, J. E., Newbury, D. E., Echlin, P., Joy, D., Fiori, C., and Lifshin, E. (1981) *Scanning Electron Microscopy and X-ray Microanalysis.* Plenum Press, New York.

Gonzalez, R. C. and Wintz, P. (1987) *Digital Image Processing* (2nd edition). Addison-Wesley, Menlo Park, Calif.

Gottlieb, P., Agron-Olshina, N., and Ho-Tun, E. (1989) The automatic measurement of mineral matter in coal. In Petruk, W., ed.,

Image Analysis in Earth Sciences. Mineralogical Association of Canada, *Short Course Handbook,* vol. 16, pp. 133–140.

Green, D. A., Denee, P. B., and Frederickson, R. G. (1979) The application of heavy metal staining (OsO_4) and backscattered electron imaging for detection of organic material in gas and oil shales. *Scanning Electron Microscopy,* 1, 475–500.

Guan, L., and Ward, R. K. (1989) Restoration of randomly blurred images by the Wiener filter. *IEEE Transactions of Acoustics, Speech, and Signal Proceedings,* 37, 589–592.

Habesch, S. M. (1990) The evaluation of pore-geometry networks in clastic reservoir lithologies using microcomputer technology. *Computers and Geology,* 6, 91–110.

Hall, M. G., and Lloyd, G. E. (1981) The SEM examination of geological samples with a semiconductor backscattered electron detector. *Am. Mineral.,* 66, 362–368.

Hall, M. G., and Lloyd, G. E. (1983) The SEM examination of geological samples with a semiconductor backscattered electron detector – reply. *Am. Mineral.,* 68, 843–844.

Heinrich, K. F. J. (1964) Interrelationships of sample composition, backscatter coefficient and target current measurement. *Adv. X-ray Anal.,* 7, 325–335.

Heinrich, K. F. J. (1966) Electron probe microanalysis by specimen current measurements. In Castaing, R., Deschamps, P., and Philbert, J., eds., *Optique des Rayons X et Microanalyses.* Hermann, Paris, 159–167.

Hejna, J. (1987) A ring scintillation detector for detection of backscattered electrons in the scanning electron microscope. *Scanning Microscopy,* 1, 983–987.

Hejna, J., Radzimski, Z., and Bukowski, A. (1985) Detection system for SEM. *Scanning Electron Microscopy,* I, 151–156.

Hounslow, M. W., and Tovey, N. K. (1992) Porosity measurement and domain segmentation of back-scattered SEM images of particulate materials. *Scanning Microscopy,* Supplement 6, 245–254.

Huggett, J. M. (1984) An SEM study of phyllosilicates in a Westphalian coal measures sandstone using backscattered electron imaging and wave length spectral analysis. *Sedimentary Geology,* 40, 233–247.

Huggett, J. M. (1986) An SEM study of phyllosilicate diagenesis in sandstones and mudstones in the Westphalian Coal Measures using back-scattered electron microscopy. *Clay Minerals,* 21, 603–616.

Jackman, J. (1980) New scanning electron microscope depends on multi-function detector. *Ind. Res. Develop.,* 22, 115–120.

Jim, C. Y. (1986) Impregnation of moist and dry unconsolidated clay samples using Spurr resin for microstructural studies. *J. Sedimentary Petrology,* 56, 596–597.

Jones, C. (1991) Characteristics and origin of rock varnish from the hyperarid coastal deserts of northern Peru. *Quat. Res.,* 35, 116–129.

Jongerius, A., and Bisdom, E. B. A. (1981) Porosity measurement using the Quantimet 720 on backscattered electron images of thin sections of soils. In Bisdom, E. B. A., ed., *Submicroscopy of Soils and Weathered Rocks.* Centre for Agricultural Publishing and Documentation, Wageningen, 207–216.

Joy, D. C., and Joy, C. S. (1993) The Micro-Channel Plate backscattered electron detector. *Microscopy and Analysis,* September, 19.

Kanaori, Y., Yairi, K., and Ishida, T. (1991) Grain boundary microcracking of granitic rocks from the northeastern region of the Atotsugawa Fault, central Japan: SEM backscattered electron images. *Eng. Geol.,* 30, 221–235.

Kimoto, S., Hashimoto, H., and Suganuma, T. (1966) Stereoscopic observation in scanning electron microscopy using multiple detectors. In McKinley, T. D., Heinrich, K. F. J., and Wittry, D. B, eds., *The Electron Microprobe.* John Wiley & Sons, New York, 480–489.

Kiss, L. T., and Briskies, H. G. (1976) Scanning electron microscopy: The use of backscattered electron image in materials investigation. *J. Aust. Inst. Min. Metall.,* 21, 178–182.

Kohler, R. (1981) A segmentation system based on thresholding. *Computer Graphics and Image Processing,* 15, 319–338.

Krinsley, D. H. (in press) Models of rock varnish formation constrained by high resolution transmission electron microscopy. *Sedimentology.*

Krinsley, D. H., and Anderson, S. (1989) Desert varnish: A new look at chemical and textural variations. *Geol. Soc. Amer. Abstr. with Programs,* 21, 103.

Krinsley, D. H., and Doornkamp, J. C. (1973) *Atlas of Quartz Sand Grain Surface Textures.* Cambridge University Press, Cambridge.

Krinsley, D. H., and Dorn, R. (1991) New eyes on eastern California rock varnish. *California Geol.,* 44, 107–115.

Krinsley, D. H., Dorn, R., and Anderson, S. (1990) Factors that may interfere with the dating of rock varnish. *Phys. Geog.,* 11, 97–119.

Krinsley, D. H., Dorn, R. I., and Tovey, K. (1995) Nanometer-scale layering in rock varnish: Implications for genesis and paleoenvironmental interpretation. *J. Geol.,* 103, 106–113.

Krinsley, D. H., and Manley, C. R. (1989) Backscattered electron microscopy as an advanced technique in petrography. *J. Geol. Educ.,* 37, 202–209.

Krinsley, D. H., Nagy, B., Dypvik, H., and Rigali, M. (1993) Micro-textures in mudrocks as revealed by backscattered electron imaging. *Precambrian Research,* 61, 191–207.

Krinsley, D. H., Pye, K., and Kearsley, A. (1983) Application of backscattered electron microscopy in shale petrology. *Geol. Mag.,* 120, 109–114.

Kuenen, Ph. H. (1960) Experimental abrasion, part 4: Eolian action. *J. Geol.,* 68, 427–449.

Kupecz, J. A., and Land, L. S. (1991) Late-stage dolomitization of the lower Ordovician Ellenburger Group, West Texas. *J. Sedimentary Petrology,* 61, 551–574.

Lanteri, H., Bindi, R., and Rostaing, P. (1988) Transport models for backscattering and transmission of low energy (<3 kilovolts) electrons from solids. *Scanning Microscopy* 2, 1927–1945.

Leckie, D., Singh, C., Goodzari, F., and Wall, J. (1990) Organic rich radioactive marine shale: A case study of a shallow-water condensed section, Cretaceous Shaftsbury Formation, Alberta, Canada. *J. Sedimentary Petrology,* 60, 101–117.

Lin, P. S. D., and Becker, R. P. (1975) Detection of backscattered electrons with high resolution. *Scanning Electron Microscopy,* 61–70.

Lipkina, M. (1990) Hydrothermal green clays in marine sediments: Clues to marine mineral deposits. *Marine Mining,* 9, 379–402.

Loendorf, L. L. (1991) Cation-ratio varnish dating and petroglyph chronology in Colorado. *Antiquity,* 65, 246–255.

Machel, H. G., Mason, R. A., Mariano, A. N., and Mucci, A. (1991) Causes and measurements of luminescence in calcite and dolomite. In Barker, E. E., and Kopp, O. C., eds., *Luminescence Microscopy and Spectroscopy: Qualitative and Quantitative Applications.* SEPM Short Course 25, Soc. Sedimentary Geology, Tulsa, Okla., pp. 9–25.

Mainwaring, P. R., and Petruk, W. (1989) Introduction to image analysis in the Earth and Mineral Sciences. In Petruk, W., ed., *Image Analysis in Earth Sciences. Mineralogical Association of Canada,* Short Course Handbook, 16, 1–5.

Mardia, K. V. (1992) *Statistics of Directional Data.* Academic Press, London.

McConnochie, I. (1974) Fabric changes in consolidated kaolin. *Geotechnique*, 24, 208–222.

Miller, J. (1988a) Microscopial techniques: I. Slices, slides, stains and peels. In Tucker, M. E., ed., *Techniques in Sedimentology*. Blackwell Scientific Publications, Oxford, 86–107.

Miller, J. (1988b) Cathodoluminescence microscopy. In Tucker, M. E., ed., *Techniques in Sedimentology*. Blackwell Scientific Publications, Oxford, 174–190.

Milliken, K. L. (1989) Petrography and composition of authigenic feldspars, Oligocene Frio Formation, South Texas. *J. Sedimentary Petrology*, 59, 361–374.

Minnis, M. M. (1984) An automatic point-counting method for mineralogical assessment. *American Assoc. Petroleum Geol. Bull.*, 68, 744–752.

Moon, C. F., and Hurst, C. W. (1984) Fabrics of muds and shales: An overview. In Stow, D. A. V., and Piper, D. J. W., eds., *Fine-Grained Sediments: Deep-Water Processes and Facies. Geol. Soc. Spec. Pub.*, 15, Blackwell, Oxford, pp. 579–593.

Morad, S., Bergan, M., Knarud, R., and Nystuen, J. P. (1990) Albitization of detrital plagioclase in Triassic reservoir sandstones from the Snorre Field, Norwegian North Sea. *J. Sedimentary Petrology*, 60, 411–425.

Morad, S., Marfil, R., and De La Peña, J. A. (1989) Diagenetic K-feldspar pseudomorphs in Triassic Buntsandstein sandstones of the Iberian Range, Spain. *Sedimentology*, 36, 635–650.

Murata, K. (1973) Monte Carlo calculations on electron scattering and secondary electron production in the SEM. *Scanning Electron Microscopy*, II, 267–276.

Murata, K. (1974) Spatial distribution of backscattered electrons in the scanning electron microscope and electron microprobe. *J. App. Phys.*, 45, 4110–4117.

Murata, K., Shimuzu, R., and Shinoda, G. (1968) Scattering of electrons in metallic targets. In Heinrich, K. F. J., ed., *Quantitative Electron Probe Microanalysis*. National Bureau of Standards Special Publication 298, Washington, D.C., 155–187.

Murphy, C. P. (1982) A comparative study of three methods of water removal prior to resin impregnation of two soils. *J. Soil Sci.*, 33, 719–735.

Nadeau, P. H., and Hurst, A. (1991) Application of back-scattered electron microscopy to the quantification of clay mineral microporosity in sandstones. *J. Sedimentary Petrology*, 61, 921–925.

Newbury, D. E., Yakowitz, H., and Mykelurst, R. L. (1973) Monte Carlo calculations of magnetic contrast from cubic materials in the scanning electron microscope. *App. Phys. Lett.,* 23, 488–490.

Nuhfer, E. (1981) Mudrock fabrics and their significance – Discussion. *J. Sedimentary Petrology,* 51, 1027–1029.

Nuhfer, E., Vinopal, R., Hohn, M., and Klanderman, D. (1981) Applications of SEM in petrology of mudrocks: Results of studies of Devonian mudrocks from West Virginia and Virginia. *Scanning Electron Microscopy,* 625–632.

O'Brien, N. (1984) Fabric of organic rich shales from various sedimentary environments. *Program Twenty-First Annual Meeting Clay Minerals Society,* Abstr., Baton Rouge, La., 90.

O'Brien, N. (1985) The role of fabric analysis in determining sedimentary environments of argillaceous rocks. *8th Int. Clay Conf.,* Abstr., Denver, Colo.

O'Brien, N. (1987) The effects of bioturbation on the fabric of shale. *J. Sedimentary Petrology,* 57, 449–455.

O'Brien , G. W., Milnes, A. R., Veeh, H. H., Heggie, D. T., Riggs, S. R., Cullen, D. J. H., Marshall, J. F., and Cook, P. J. (1990) Sedimentation dynamics and redox iron cycling: Controlling factors for the apatite-glauconite association on the East Australian continental margin. In Notholt, A. J., and Jarvis, I., eds., *Phosphorite Research and Development,* Geological Society Special Publication, No. 52, London, 61–86.

O'Brien, N., and Slatt, R. M. (1990) *Argillaceous Rock Atlas.* Springer-Verlag, New York.

Odin, G., ed. (1988) Green Marine Clays. *Developments in Sedimentology 45.* Elsevier, Amsterdam.

Odin, G. S., and Dodson, M. H. (1982) Zero isotopic ages of glauconites. In Odin, G. S., ed., *Numerical Dating in Stratigraphy.* John Wiley & Sons, New York, 277–306.

Odin, G. S., and Matter, A. (1981) De glauconiarum origine. *Sedimentology,* 28, 611–642.

Odom, I. E. (1976) Microstructure, mineralogy and chemistry of Cambrian glauconite pellets and glauconite, central U.S.A. *Clays and Clay Minerals,* 24, 232–238.

Odom, I. E. (1984) Glauconite and celadonite minerals. In Bailey, S. W., ed., *Micas. Reviews in Mineralogy,* Mineralogical Society of America, 13, 545–572.

Orford, J. D., and Whalley, W. B. (1983) The use of fractal dimensions to quantify the morphology of irregular shaped particles. *Sedimentology,* 30, 655–668.

Orford, J. D., and Whalley, W. B. (1991) Quantitative grain form analysis. In Syvitski, J., ed., *Principles, Methods and Application of Particle Size Analysis.* Cambridge University Press, Cambridge, pp. 88–108.

Palmer, F. E., Staley, J. T., Murray, R. G., Counsell, T., and Adams, J. B. (1985) Identification of manganese-oxidizing bacteria from desert varnish. *Geomicrobiology Journal,* 4, 343–360.

Petruk, W. (1989a) Image analysis of minerals. In Petruk, W., ed., *Image Analysis in Earth Sciences.* Mineralogical Association of Canada, *Short Course Handbook,* 16, 6–18.

Petruk, W. (1989b) Image analysis and mineral beneficiation. In Petruk, W., ed., *Image Analysis in Earth Sciences.* Mineralogical Association of Canada, *Short Course Handbook,* 16, 86–89.

Petruk, W., ed. (1989c) *Image Analysis in Earth Sciences.* Mineralogical Association of Canada, *Short Course Handbook,* 16.

Pineda, C. A., Peisach, M., Jacobson, L., and Sampson, C. G. (1990) Cation-ratio differences in rock patina on hornfels and chalcedony using Thick Target PIXIE. *Nuclear Instruments and Methods in Physics Research,* B49, 332–335.

Potter, P., Maynard, J., and Pryor, W. A. (1980) *Sedimentology of Shale.* Springer-Verlag, New York.

Potter, R., and Rossman, G. (1977) Desert varnish: The importance of clay minerals. *Science,* 196, 1446–1448.

Potter, R., and Rossman, G. (1979) The manganese and iron mineralogy of desert varnish. *Chem. Geol.,* 25, 79–94.

Prior, D., and Behrmann, J. (1989) Backscatter imagery of fine-grained sediments from Hole 671B, DSDP Leg 110: Preliminary results. *Proceedings of the Ocean Drilling Program,* vol. 110, Part B. U.S. Government Printing Office, Washington, D.C., 245–255.

Prod'homme, M., Coster, M., Chermant, L., Chermant, J.-L. (1992) Morphological filtering and granulometric analysis on scanning electron micrographs. *Applications in Material Science, Scanning Microscopy,* Supplement 6, 255–268.

Protz, R., Sweeney, S. J., and Fox, C. A. (1992) An application of spectral image analysis to soil micromorphology, I: Methods of analysis. *Geoderma,* 53, 275–288.

Purvis, K. (1989) Zoned authigenic magnesites in the Rotliegend Lower Permian, southern North Sea. *Sedimentary Geology,* 65, 307–318.

Pye, K. (1984a) Rapid estimation of porosity and mineral abundance in backscattered electron images using a simple SEM image analyzer. *Geol. Mag.,* 121, 81–84.

Pye, K. (1984b) SEM analysis of siderite cements in intertidal marsh sediments, Norfolk, England. *Mar. Geol.,* 56, 1–12.

Pye, K. (1985a) Electron microscope analysis of zoned dolomite rhombs in the Jet Rock Formation (Lower Toarcian) of the Whitby area, U. K. *Geol. Mag.,* 122, 279–286.

Pye, K. (1985b) Granular disintegration of gneiss and migmatites. *Catena,* 12, 191–199.

Pye, K., and Coleman, M. L. (in press) Texture, mineralogy and geochemistry of Quaternary dolocretes and dolocalcretes from the Kora area of central Kenya. *Sedimentary Geology.*

Pye, K., Dickson, J. A. D., Schiavon, N., Coleman, M. L., and Cox, M. L. (1990) Formation of siderite-Mg calcite-iron monosulfide concretions in intertidal marsh and sandflat sediments, north Norfolk, England. *Sedimentology,* 37, 325–343.

Pye, K., and Krinsley, D. H. (1983) Mudrocks examined by backscattered electron microscopy. *Nature,* 301, 412–413.

Pye, K., and Krinsley, D. (1984) Petrographic examination of sedimentary rocks in the SEM using backscattered electron detectors. *J. Sedimentary Petrology,* 54, 877–888.

Pye, K., and Krinsley, D. H. (1985) Formation of secondary porosity in sandstones by quartz framework grain dissolution. *Nature,* 317, 54–56.

Pye, K., and Krinsley, D. H. (1986a) Diagenetic carbonate and evaporite minerals in Rotliegend aeolian sandstones of the southern North Sea: Their nature and relationship to secondary porosity development. *Clay Minerals,* 21, 443–457.

Pye, K., and Krinsley, D. (1986b) Microfabric, mineralogy and early diagenetic history of the Whitby Mudstone Formation (Toarcian), Cleveland Basin, U. K. *Geol. Mag.,* 123, 191–203.

Pye, K., Krinsley, D., and Burton, J. (1986) Diagenesis of U. S. A. Gulf Coast shales reconsidered. *Nature,* 324, 557–559.

Pye, K., and Mazzullo, J. M. (1994) Effects of tropical weathering on quartz grain shape: An example from Northeastern Australia. *J. Sedimentary Petrology,* A 64, 500–507.

Pye, K., and Miller, J. A. (1988) Chemical and biochemical weathering of pyritic mudrocks in a shale embankment. *Q. J. Eng. Geol.,* 23, 365–381.

Pye, K., and Mottershead, D. N. (1995) Honeycomb weathering of Carboniferous sandstone in a sea wall at Weston-super-Mare, U.K. *Q. J. Eng. Geol.,* 28, 333–347.

Pye, K., and Windsor-Martin, J. (1983) SEM analysis of shales and other geological materials using the Philips Multi-Function Detector system. *The EDAX Editor,* 13, 4–6.

Radzimski, Z. J. (1987) Scanning electron microscope solid state detectors. *Scanning Microscopy,* 1, 975–982.

Razaz, M., Lee, R., and Shaw, P. (1993) A nonlinear iterative least-squares algorithm for image restoration. Proceedings IEEE, *Nonlinear Signal Processing,* 4.1–4.6.

Reed, S. J. B. (1975) *Electron Microprobe Analysis.* Cambridge University Press, Cambridge.

Reeder, R. J., and Prosky, J. L. (1986) Compositional sector zoning in dolomite. *J. Sedimentary Petrology,* 56, 237–247.

Reimer, L., and Riepenhausen, M. (1985) Detector strategy for secondary and backscattered electrons using multiple detector systems. *Scanning* 7, 221–238.

Reimer, L., and Volbert, B. (1980) Advantages of two opposite Everhart-Thornley detectors in SEM. *Scanning Electron Microscopy,* IV, 1–10.

Reneau, S. L., Raymond, R., and Harrington, C. D. (1992) Elemental relationships in rock varnish stratigraphic layers, Cima Volcanic Field, California: Implications for varnish and the interpretation of varnish chemistry. *American Journal of Science,* 292, 684– 723.

Robinson, B. W., and Nickel, E. (1979) A useful new technique for mineralogy: The BSE/low vacuum mode of SEM operation. *Am. Mineral.,* 64, 1322–1328.

Robinson, B. W., and Nickel, E. H. (1983) The SEM examination of geological samples with a semiconductor backscattered electron detector: A discussion. *Am. Mineral.,* 68, 840–842.

Robinson, V. N. E. (1973) A reappraisal of the complete electron emission spectrum in scanning electron microscopy. *J. Phys. D.,* 6, L105–L107.

Robinson, V. N. E. (1975) Backscattered electron imaging. *Scanning Electron Microscopy,* 51–60.

Robinson, V. N. E. (1980) Imaging with backscattered electrons in a scanning electron microscope. *Scanning,* 3, 15–26.

Robinson, V. N. E. (1988) Backscattered electron imaging with a scintillator detector. *Microscopy and Analysis,* March, 7–9.

Robinson, V. N. E., and George, E. P. (1978) Electron scattering in the SEM. *Scanning Electron Microscopy,* I, 859–866.

Ruzyla, K. (1986) Characterization of pore space by quantitative image analysis. *SPE Formation Evaluation,* August, 386–389.

Saigal, G. C., Morad, S., Bjørlykke, K., Egeberg, P. K., and Aagaard, P. (1988) Diagenetic albitization of detrital K-feldspar in Jurassic, Lower Cretaceous, and Tertiary Clastic Reservoir rocks from offshore Norway, I: Textures and origin. *J. Sedimentary Petrology,* 58, 1003–1013.

Saparin, G. V., and Obyden, S. K. (1993) Colour in the microworld: Real colour cathodoluminescence mode in scanning electron microscopy. *Microscopy and Analysis*, March, 5–7.

Sawyer, G. R., and Page, T. F. (1978) Microstructural characterization of "REFEL" (reduction bonded) silicon carbides. *J. Mat. Sci.*, 13, 885–904.

Schiavon, N. (1992) BSEM study of decay mechanisms in urban oolitic limestones. *European Cultural Heritage Newsletter*, 6(3), 35–46.

Schiavon, N., Chiavari, G., Fabbri, D. & Schiavon, G. (1994) Microscopical and chemical analysis of black patinas on granite. In Fassina, G., Ott, H., and Zezza, F., eds., *3rd International Symposium on the Conservation of Monuments in the Mediterranean Basin, Venice, 22–25 June*, 93–99.

Schieber, J. (1986) The possible role of benthic microbial mats during the formation of carbonaceous shales in shallow Mid-Proterozoic basins. *Sedimentology*, 33, 521–536.

Schieber, J. (1989) Facies and origin of shales from the Mid-Proterozoic Newland Formation, Belt Basin, Montana, U.S.A. *Sedimentology*, 36, 203–219.

Schieber, J. (1994) Evidence for high-energy events and shallow-water deposition in the Chattanooga Shale, Devonian, central Tennessee, U.S.A. *Sedimentary Geology*, 93, 193–208.

Schur, K., Blaschke, R., and Pfefferkorn, G. (1974) Improved conditions for backscattered electron SEM micrographs on polished sections using a modified scintillator detector. *Scanning Electron Microscopy*, 1003–1010.

Seyedolali, A., and S. Boggs, Jr. (1996) Albitization of Miocene deep-sea sandstones from the Japan Sea backarc basin. *J. Sed. Soc. Japan*, 43, 1–18.

Shehata, M. T. (1989) Applications of image analysis in characterizing dispersion of particles. In Petruk, W., ed., *Image Analysis in Earth Sciences*, Mineralogical Association of Canada, *Short Course Handbook*, 16, 119–132.

Smart, P., and Leng, X. (1993) Present developments in image analysis. *Scanning Microscopy*, 7, 5–16.

Smart, P., and Tovey, N. K. (1981–1982) *Electron Microscopy of Soils and Sediments Volume I*. Techniques. Oxford University Press, Oxford.

Smart, P., and Tovey, N. K. (1988) Theoretical Aspects of Intensity Gradient Analysis. *Scanning*, 10, 115–121.

Smart, P., Tovey, N. K., McConnochie, I., Leng, X., and Hounslow, M. W. (1990) Automatic analysis of electron microstructure of cohesive sediments. In Bennett, R. H., Bryant, W. R., and Hulbert, M. H., eds., *Microstructure of Fine-Grained Sediments from Mud to Shale*, Springer-Verlag, New York, 359–366.

Smith, M. M. (1989) Shape analysis of loose particles: From shape factor to fractals to Fourier. In Petruk, W., ed., *Image Analysis in Earth Sciences. Mineralogical Association of Canada*, Short Course Handbook, 16, 106–118.

Sokolov, V. N., (1990) Engineering–geological classification of clay microstructure. In Price, D. G., ed., Proc. 6th Int. Congr., *International Association of Engineering Geology*, Balkema, Rotterdam, 753–760.

Sorby, H. C. (1877a) The application of the microscope to geology. *Month. Micros. J.,* 17, 113–136.

Sorby, H. C. (1877b) On a new method for determining the index of refraction of small portions of transparent materials. *Min. Mag.,* 6, 193–208.

Sorby, H. C. (1878) On the determination of the minerals in thin sections of rocks by means of their indices of refraction. *Min. Mag.,* 8, 1–4.

Spurr, A. R. (1969) A low viscosity epoxy resin embedding medium for electron microscopy. *J. Ultrastructure Res.,* 26, 31–43.

Stephen, J., Smith, B. J., Marshall, D. C., and Wittam, E. M. (1975) Applications of semiconductor backscattered electron detector in a scanning electron microscope. *J. Phys. E.,* 8, 607–618.

Stuart, C. J., Vavra, C. L., and Levesque, T. H. J. (1987) Analysis of microporosity in reservoir rocks by combined scanning electron microscopy and digital image analysis. *Microbeam Analysis,* 327–328.

Surdam, R. C., Crossey, L. J., Hagen, E. S., and Heasler, H. P. (1989) Organic-inorganic interactions and sandstone diagenesis. *Am. Assoc. Petroleum Geologists Bull.,* 73, 1–23.

Taggart, J. E. Jr. (1977) Polishing technique for geologic samples. *Am. Mineral.,* 62, 824–827.

Terribile, F., and Fitzpatrick, E. A. (1995) The application of some image analysis techniques to the recognition of soil micromorphological features. *European Journal of Soil Science,* 46, 29–45.

Tovey, N. K. (1971) Soil structure analysis using optical techniques on scanning electron micrographs. In Johari, O., ed., *Proc. 4th Int Symposium of Scanning Electron Microscopy.* IIT Research Institute, Chicago, 49–56.

Tovey, N. K. (1980) A digital computer technique for orientation analysis of micrographs of soil fabric. *Journal of Microscopy,* 120, 303–315.

187

Tovey, N. K. (1995) Techniques to examine the microfabric and particle interactions of collapsible soils. In Derbyshire, E., Dijkstra, T., and Smalley, I. J., eds., *Genesis and Properties of Collapsible Soils*. Kluwer, Dordrecht, The Netherlands, pp. 65–92.

Tovey, N. K., and Dent, D. L. (in press) A gray-level morphological algorithm for analysing soil micro fabrics. *Proc. 10th Int. Working Meeting on Soil Micromorphology*. Moscow.

Tovey, N. K., Dent, D. L., Krinsley, D. H., and Corbett, W. M. (1994a) Quantitative micromineralogy and microfabric of soils and sediments. In Ringrose-Voase, A. J., and Humphreys, G. S., eds., *Soil Micromorphology*. Elsevier, Amsterdam, 541–547.

Tovey, N. K., Dent, D. L., Krinsley, D. H., and Corbett, W. M. (1992a) Processing multispectral SEM images for quantitative microfabric analysis. *Scanning Microscopy*, Supplement 6, 269–282.

Tovey, N. K., and Hounslow, M. W. (1995) Quantitative microporosity and orientation analysis in soils and sediments. *J. Geol. Soc. Lond.*, 152, 119–129.

Tovey, N. K., Hounslow, M. H., and Wang, J.-M. (1995) Orientation analysis and its application in image analysis. *Advances in Imaging & Electron Physics*, 93, 219–329.

Tovey, N. K., Krinsley, D. H, Dent, D. L., and Corbett, W. M. (1992b) Techniques to quantitatively study the microfabric of soils. *Geoderma*, 53, 217–235.

Tovey, N. K., and Krinsley, D. H. (1991) Mineralogical mapping of scanning electron micrographs. *J. Sedimentary Petrology*, 75, 109–123.

Tovey, N. K., and Krinsley, D. H. (1992) Mapping the orientation of fine-grained minerals in soils and sediments. *Bull. Int. Ass. Eng. Geol.*, 46, 93–101.

Tovey, N. K., and Martinez, M. D. (1991) A comparison of different formulae for orientation analysis of electron micrographs. *Scanning*, 13, 289–298.

Tovey, N. K., and Smart, P. (1986) Intensity gradient techniques for orientation analysis of electron micrographs. *Scanning*, 8, 75–90.

Tovey, N. K., Smart, P., and Hounslow, M. W. (1990) Quantitative orientation analysis of soil microfabric. In Douglas, L. A., ed., *Soil micromorphology: A Basic and Applied Science*. Elsevier, Amsterdam, 631–639.

Tovey, N. K., Smart, P., and Hounslow, M. W. (1994b) Quantitative methods to determine micro-porosity in soils and sediments. In Ringrose-Voase, A. J., and Humphreys, G. S., eds., *Soil Micromorphology*. Elsevier, Amsterdam, 531–539.

188

Tovey, N. K., Smart, P., Hounslow, M. W., and Desty, J. P. (1992c) Automatic orientation analysis of microfabric. *Scanning Microscopy*, Supplement 6, 315–330.

Tovey, N. K., Smart, P., Hounslow, M. W., and Leng, X. L. (1989) Practical aspects of automatic orientation analysis of micrographs. *Scanning Microscopy*, 3, 771–784.

Tovey N. K., Smart, P., Hounslow, M. W., and Leng, X. L. (1992d) Automatic mapping of some types of soil fabric. *Geoderma*, 53, 179–200.

Tovey, N. K., and Sokolov, V. N. (1981) Quantitative methods for soil fabric analysis. *Scanning Electron Microscopy*, Series I, 536–554.

Tovey, N. K., and Sokolov, V. N. (in press) Image analysis applications in soil micromorphology. *Proc. 10th Int. Working Meeting on Soil Micromorphology*. Moscow.

Tovey, N. K., and Wong, K. Y. (1974) Some aspects of quantitative measurements from electron micrographs of soil structure. In Rutherford, G. K., ed., *Soil Microscopy*. Kingston, Ontario, Limestone Press, pp. 207–222.

Tovey, N. K., and Wong, K. Y. (1978) Optical techniques for analysis scanning electron micrographs. *Scanning Electron Microscopy*, 1, 381–392.

Trewin, N. (1988) Use of scanning electron microscope in sedimentology. In Tucker, M., ed., *Techniques in Sedimentology*. Blackwell Scientific Publications, Oxford, 229–273.

Unitt, B. M. (1975) A digital computer method for revealing orientation information in images. *J. of Phys. E. Series 2*, 8, 423–425.

Van Houten, F. B., and Purucker, M. E. (1984) Glauconitic peloids and chamositic ooids – favorable constraints and problems. *Earth Science Reviews*, 20, 211–243.

Volbert, B. (1982) Signal mixing techniques in scanning electron microscopy. *Scanning Electron Microscopy*, III, 897–906.

Weaver, C. (1989) *Clays, Muds and Shales*. Elsevier, Amsterdam.

Wells, O. C. (1970) New contrast mechanism for SEM. *App. Phys. Lett.*, 16, 151–153.

Wells, O. C. (1977) Backscattered electron image in the scanning electron microscope. *Scanning Electron Microscopy*, I, 747–771.

Wells, O. C. (1978) Penetration effects at sharp edges in the scanning electron microscope. *Scanning* 1, 58–62.

Wells, O. C. (1986) Reduction of edge penetration effect in the SEM. *Scanning*, 8, 120–126.

Welton, J. E. (1984) *SEM Petrology Atlas*. American Association of Petroleum Geologists, Tulsa, Okla.

White, S. H., Huggett, J. M., and Shaw, H. F. (1985) Electron-optical studies of phyllosilicate intergrowths in sedimentary and metamorphic rocks. *Mineralogical Magazine*, 49, 413–423.

White, S. H., Shaw, H. F., and Huggett, J. M. (1984) The use of backscattered electron imaging for petrographic study of sandstones and shales. *J. Sedimentary Petrology*, 54, 487–494.

Wittry, D. B. (1966) Secondary electron emission in the electron probe. In Castaing, R., Deschamps, P., and Philbert, J., eds., *Optique des Rayons X et Microanalyses.* Hermann, Paris, 168–178.

Wolf, E. D., and Everhart, T. E. (1969) Annular diode detector for high angular resolution pseudo-kikuchi patterns. *Scanning Electron Microscopy,* 43–44.

Xie, G., and Shen, P. (1991) A discovery and primary study of glauconite in the Upper Triassic Yanchang oil-bearing sandstone in northern Shaanxi. *Scientia Geologica Sinica,* 126–136. (Chinese, with English abstract)

Zempolich, W. G., and Baker, P. A. (1993) Experimental and natural mimetic dolomitization of aragonite ooids. *J. Sedimentary Petrology,* 63, 596–606.

Zuniga, O. A., and Haralick, R. M. (1987) Integrated directional derivative gradient operator. *IEEE Transactions*, SMC-17, 508–517.

Index

albite, 10, 34, 61–5, 67–70, 81, 141, 142
algal tubules, 104, 106
aluminum, 163–5
ammonite, 108, 109
amphiboles, 29
anhydrite, 57, 59, 67, 70–2, 118
ankerite, 98, 103–5
apatite, 118
aragonite, 43, 45, 85, 98, 107, 113
atomic number, 5, 8, 10, 12, 15, 16, 25, 34, 53, 117, 170
automated mineral identification, 146, 170

backscattered electrons, 4–9, 11, 13, 14, 15, 17–20, 123
bacteria, 121, 124, 125, 127
barite, 117, 118, 123
binary segmentation, 152, 154, 155, 157
biotite, 10, 34, 56, 79

calcite, 34, 41, 43, 46, 47, 48, 50, 59, 63, 78, 82, 85–8, 91, 98, 103–5, 108, 110, 111, 114, 115, 117, 138, 167
calcite cement, 104, 106, 107, 108
calcium, 103, 164, 165
calcium carbonate, 31, 57, 52, 58, 70, 71, 73, 76, 84, 88, 89, 90
carbonate cement, 104, 106,109
carbonate grains, 98, 99

cathodoluminescence, 5, 99, 100, 101, 104, 107, 108, 146, 162
cement stratigraphy, 104, 106, 107, 108
chert, 110, 118
chlorite, 34, 55, 60, 61, 80, 81, 84, 87, 88, 89, 92, 93, 142, 143
clasts, 98
clay minerals, 48, 52, 55, 59, 70, 73, 76, 77, 79, 81, 83, 85, 88, 89, 90, 92, 94, 95, 119, 121–3, 127, 132, 135, 138, 142, 163, 165
coral, 107
cryptalgal fabrics, 99

diagenesis, 25–52, 55, 57, 61, 65, 67, 68, 70–2, 75, 78–82, 84–9, 91, 92, 99, 104, 113–16, 121, 123, 129, 139, 143
digital images, 146–51, 154
dolomite, 10, 34, 43, 46–8, 57–9, 71, 72, 89, 98–105, 110, 112, 113, 117, 118
domain-segmented images, 159–61, 168, 170
duricrust, 43

evaporites, 48, 57, 71

fabric orientation, 145, 146
feldspar, 29, 37, 46, 48, 49, 52, 54, 55, 59, 60, 65, 73, 77, 83, 85, 86, 88, 89, 90, 117, 121, 126, 131–3, 141–3, 163

foraminifer, 85–8, 103–6, 110–12, 116, 134, 135, 137–9
fossils, 84, 85, 89, 98, 104, 115, 124, 131, 132, 133, 137, 138, 139, 140
fractures, 127, 128, 133, 134, 135, 136, 137, 138, 139, 141, 143
fossil ghosts, 137, 139

glauconite, 77, 89, 95, 131–44
grain shape, 26, 27, 65, 74, 81, 83, 97, 145, 146, 154, 155, 162, 169, 170, 171
grain size, 26, 27, 65, 73, 74, 81, 83, 97, 145, 146, 154, 155, 162, 169, 170, 171
gray level, 26, 37, 53, 54, 62, 68, 100, 147–9, 152, 154, 155, 157, 163, 169, 170, 171
gypsum, 31, 34, 49, 91

halite, 32
hematite, 10, 57, 59, 100, 118

illite, 10, 25, 29, 55–7, 60, 61, 67, 71, 72, 77, 79, 80, 81, 84, 87–9, 92, 95, 121, 142, 143
ilmenite, 26, 29
image acquisition, 146
image analysis, 26, 52, 53, 65, 66, 97, 145–72
image classification, 165–8
image processing, 146, 151–4, 157, 168
intensity-gradient analysis, 157
iron, 100–103, 107, 108, 109, 132, 135, 136, 163, 165
iron oxide, 110, 142, 143

k-feldspar, 29, 30, 34, 48, 60, 70, 165, 167
kaolinite, 10, 28, 29, 34, 55–8, 60, 77–81, 84–9, 92, 93, 121, 142

laminae, 75, 76, 77, 78, 81, 93, 113, 119, 121, 123, 124, 129

magnesian calcite, 40, 98, 99, 103, 108, 110

magnesite, 98
magnesium, 100, 110, 111, 163
manganese, 100, 107–9
manganese nodules, 36, 37, 38
mica, 29, 30, 46, 73, 76, 78, 79, 80, 88, 92, 93, 117, 118, 121, 131, 141–4
micrite, 115
morphological analysis, 168
multispectral methods, 157, 158, 162–6, 168, 170
muscovite, 34, 55, 78, 79

nannofossils, 111
neomorphic calcite, 113

ooids, 98, 113, 114, 137, 141, 143
orthoclase, 10

particle orientation, 146, 155, 156, 158, 159, 169, 171
peloids, 98
pixels, 147, 148, 151–6, 158, 162, 163, 165, 166, 168, 170
plagioclase, 29, 30, 54, 61–5, 68, 70, 141–3
polished specimens, 12, 21
porosity, 39, 40, 52, 55–9, 64–8, 70–2, 82, 85, 92, 95–7, 99, 101,104, 106, 112, 113, 115–17, 127, 129, 145, 146, 154, 155, 168–71
potassium, 135, 136, 163–5
pyrite, 10, 31, 52, 76, 78, 82–94, 106, 109, 113, 118, 143, 163, 165, 167

quartz, 10, 12, 25, 28–30, 34, 39, 46, 48–61, 63, 66, 67, 71–3, 76–8, 80–91, 103, 104, 117, 121, 123, 126, 141–3, 163, 165, 167

rock fragments, 54, 59, 141
rutile, 29

secondary electrons, 4, 5, 11–3, 15, 16, 17, 26, 123

siderite, 10, 37, 39, 40–5, 57, 59, 71, 72, 90, 98
silicon, 163, 164
smectite, 35, 46, 88, 89, 92, 95, 121, 131
sparry calcite cement, 98, 99
specimen preparation, 21–5, 65, 97
stromatactis, 99
stromatolites, 126, 127
strontionite, 98
strontium, 110, 111
stylolites, 113
sulfates, 107, 163, 165
syntaxial rims, 100

training areas, 163, 165, 166

varnish textures, 119, 120, 123, 127

weathering, 27, 28, 29, 31, 32, 33, 46, 52

x-ray element maps, 146, 148, 162, 163, 165

zeolites, 36, 59, 62, 73
zircon, 25, 26
zoned carbonate cement, 99
zoned calcite, 104, 105, 108, 109
zoned dolomite, 99–105